How Animals Have Sex

How Animals Have Sex

GIDEON DEFOE

Weidenfeld & Nicolson
LONDON

First published in Great Britain in 2005
by Weidenfeld & Nicolson

10 9 8 7 6 5 4 3 2 1

The right of Gideon Defoe to be identified as the author of
this work has been asserted by him in accordance with the
Copyright, Designs and Patents Act 1988.

A CIP catalogue record for this book is available
from the British Library

ISBN 0 297 85112 8 (Hardback)
     0 297 85242 6 (Trade Paperback)

Printed and bound in Italy

Weidenfeld & Nicolson
An imprint of the Orion Publishing Group
Orion House, 5 Upper St Martin's Lane, London WC2H 9EA

www.orionbooks.co.uk

DEDICATION

For the female bean weevil,
who has a miserable time,
and probably needs some cheering up.

Contents

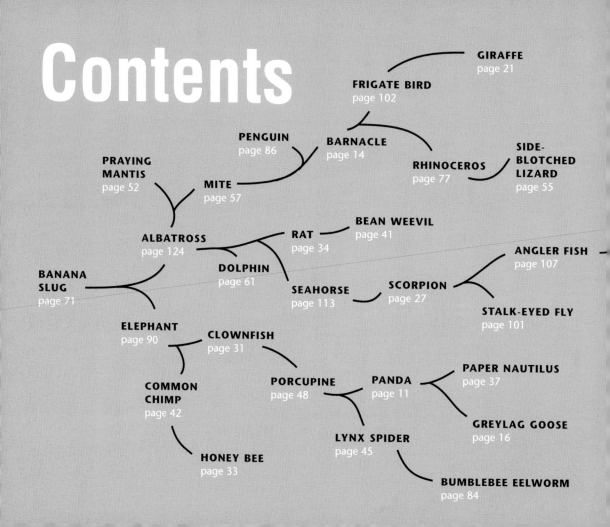

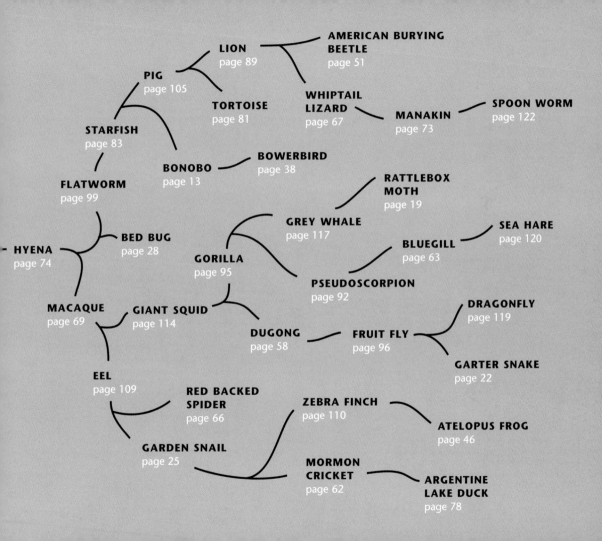

LION
page 89

AMERICAN BURYING BEETLE
page 51

PIG
page 105

WHIPTAIL LIZARD
page 67

TORTOISE
page 81

MANAKIN
page 73

SPOON WORM
page 122

STARFISH
page 83

BOWERBIRD
page 38

BONOBO
page 13

FLATWORM
page 99

RATTLEBOX MOTH
page 19

GREY WHALE
page 117

SEA HARE
page 120

BLUEGILL
page 63

HYENA
page 74

BED BUG
page 28

GORILLA
page 95

PSEUDOSCORPION
page 92

DRAGONFLY
page 119

MACAQUE
page 69

GIANT SQUID
page 114

DUGONG
page 58

FRUIT FLY
page 96

GARTER SNAKE
page 22

EEL
page 109

RED BACKED SPIDER
page 66

ZEBRA FINCH
page 110

ATELOPUS FROG
page 46

GARDEN SNAIL
page 25

MORMON CRICKET
page 62

ARGENTINE LAKE DUCK
page 78

Introduction

I cannot promise that this book is going to teach you any fundamental truths about evolutionary theory. Nor will it use the sociobiology of other creatures to throw any light on human sexuality. Animals do stupid things. Humans do stupid things. All you'll really learn is that everything in the world is almost equally stupid. If you're more than about eight years old, you already know this. And if you're less than eight years old, you shouldn't be reading books with 'animal sex' in the title.

Despite all that there are still a **whole raft of reasons** why this book will prove an invaluable aid to you. For example:

- **Anecdotes for social occasions**. Especially when you're trying to turn the conversation towards romance. A good way to do this, if you happen to be walking the object of your affections home under a starry sky, might be to say, 'What a beautiful night. Look, there's Ursa Major, or the Great Bear. Bears don't really do anything that interesting sexually, but did you know that some species of nematode have genitals that grow to hundreds of times their original size?' The mood is automatically set for love.
- The world of reproductive strategies is a fantastic and mostly untapped source of new insults, ie 'You ate all the biscuits, **you sperm-vomiting, dwarf-male spoon worm!**' And so on.
- You might go to Space for what feels to you like just a couple of weeks, only to discover upon your return that millions of years have passed back on Earth, and that **hyenas have become the new dominant species**. Thanks to this book, you'll know that the best way to greet your new masters is by touching penises for a few minutes.
- At a push you could make out that reading about animal sex shows you **care about the environment**. I'm not exactly sure how, something to do with a lot of animals living in rainforests.

But the main reason you should read this is that I don't see why I should have to know all these terrible, terrible things and you should get off scot free.

GIANT PANDA

SCIENTIFIC NAME..*Ailuropoda melanoleuca*
CLASSIFICATION...Phylum: Chordata, Class: Mammalia, Order: Carnivora, Family: Ursidae
SIZE.............140-150 cm, 90-140 kg
DISTRIBUTION.....Central China
DIET.............Bamboo, insects

When 99% of your diet is bamboo – which has slightly less nutritional value than a Happy Meal – it's not really surprising that you can't get up enough energy to start winking suggestively at other pandas. Not that you would have time to do anything about it if one of them happened to wink back anyhow, because you have to spend a massive two-thirds of your life filling your face just to stay awake. Then there's the problem that female pandas can only conceive for two or three days in any given year. As everybody knows, the odds are stacked pretty high against panda procreation.

But luckily pandas have one extremely useful, though unintentional, adaptation. Because they need such huge jaws to be able to get through all that tough-as-old-boots bamboo, pandas are blessed with **ridiculously outsized heads**. This makes them look like **giant lovable babies** to us humans, and as a result we go to daft lengths to try to help them produce as many little pandas as possible. If pandas looked like dirty great earwigs instead of doe-eyed teddy bears constantly giving us a cheery thumbs up, they'd be in a lot of trouble.

The Chinese government has an added incentive to encourage the pandas to get it together, as they pocket a million dollars per year for every panda loaned out to a foreign zoo. So at the Chengdu Giant Panda Breeding and Research Centre they hit on the idea of sitting pandas down in front of a telly, turning down the lights and showing them steamy videos of other pandas having **amazing panda sex**. It's not clear from the published articles where they got this original film from in the first place. But regardless of whether it was actually just a couple of unlucky political dissidents being made to dress up in panda suits, in at least two cases the videos seemed to do the trick and lead to successful copulation. Obviously there's no way of knowing whether, after watching all that porn, either Didi or Ximeng were consumed by a terrible sense of self-loathing and hollow regret. But really, who cares? Look at their adorable big heads!

BONOBO (PYGMY CHIMPANZEE)

SCIENTIFIC NAME ..*Pan paniscus*
CLASSIFICATION...Phylum: Chordata, Class: Mammalia, Order: Primates,
                Family: Hominidae
SIZE.............70-80 cm, 27-61 kg
DISTRIBUTION.....The Congo river basin
DIET............Fruits, seeds, insects (rarely)

For years it was thought only humans engaged in face-to-face sex. This was actually held up as a great example of why we were so much better than the rest of the animal kingdom. All that staring-meaningfully-into-each-others'-eyes business apparently proved how deep we were. We even sent missionaries around the world to make sure everybody was using the correct civilised position and weren't accidentally doing it like beasts.

But it turns out there is at least one other creature that enjoys frontal intercourse – our closest cousin, the bonobo. They are also the only other animal we know of who use tongues whilst kissing. And they do it a lot.

It's not all just for fun or procreation: sex plays an important role in resolving disputes and maintaining social cohesion. Put simply, bonobo societies get along brilliantly because they're **always** getting it on with each other.[1] In any possible combination. Girls carry other girls about on their backs and rub clitorises together. Boys find other boys and rub their rumps together. They happily engage in fellatio and mutual masturbation all over the place. And they're not selfish about it either – during sex the male bonobo will alter the speed and intensity of his thrusting based on the facial expression and vocalisations of the female. Whereas other animals get into fights over food or territory, **bonobos just have sex instead**. Imagine a world where all your petty little disputes are solved like that. 'Oh, I'm sorry, Jennifer Garner, I appear to have eaten your termites. Listen, I'd hate for there to be any bad feeling between us. So...'

I think we could all learn a lot from the bonobo way of doing things. Except for the part about publicly masturbating in front of tourists. Best ignore that bit.

[1] Primatologist Frans de Waal cautions us that '[the bonobo's] sexiness should not be exaggerated. Bonobos do not, in fact, enagage in sex all the time. At the zoo, the average bonobo initiates sex once every one and a half hours.' I suspect Frans is just showing off here by trying to pretend he thinks once every one and a half hours is nothing to write home about.

BARNACLE

```
SCIENTIFIC NAME..Balanus crenatus
CLASSIFICATION...Phylum: Arthropoda, Class: Crustacea, Order: Thoracica,
                 Family: Balanic
SIZE............Up to 25 mm
DISTRIBUTION.....Inter-tidal zone to deep sea
DIET............Plankton, gametes
```

There are worse things than being a barnacle. Stuck to the bottom of pirate boats, they get to visit exotic locations, dig up gold, meet Spanish princesses and take part in exciting sea battles. Bad enough that this tiny crustacean's life is already more glamorous than yours, but it gets even worse.

Barnacles are held fast to whatever they happen to be sat on by a protein polymer they exude that's as strong as the glue used to hold a space shuttle together. So they're faced with an obvious problem – how do they have sex with another barnacle when none of them can move? Their solution isn't particularly ingenious, but it is impressive – the barnacle has evolved a penis that is **up to thirty times as long as the rest of its body**. That's over one hundred and eighty feet if barnacles were scaled up to human size. Or, to use proper scientific textbook terminology:

> Six double-decker buses stacked on top of each other
> Three and a half Tyrannosaurus rexes
> A couple of St Paul's cathedrals.

Where you have a cluster of barnacles, they'll often all have sex with each other, and because there are so many, they quickly get to the point where none of them can hold any more barnacle sperm, so it starts to leak out and they end up covered in the stuff.

There are no actual barnacles the size of a person, which is probably a good thing.

GREYLAG GOOSE

SCIENTIFIC NAME..*Anser anser*
CLASSIFICATION...Phylum: Chordata, Class: Aves, Order: Anseriformes, Family: Anatidae
SIZE.............3.5-6.5 kg, 55-100 cm
DISTRIBUTION.....Europe, Siberia, Central Asia
DIET.............Aquatic plants

MICHAEL GRADE, BBC

Dear sir,

Still haven't heard back from you regarding my proposed situation comedy *Stone the Crows!* about two noisy crows living above an angry bear. I am putting this down to our Third World postal service. Anyhow, you'll be pleased to hear that I have subsequently had another fantastic idea for a brilliant **creature-based sitcom**.

I don't know if you have much call to read up on geese population dynamics in your line of work, but you may be aware of the extremely high incidence of homosexuality amongst certain breeds. In some species of Greylag goose, up to 14% have been observed to form homosexual couples. Sometimes a third goose, a female, will try to work her way in between the two boys, and get herself impregnated. This ends up as a happy little threesome, with all of the birds helping to raise their young. You can imagine how this unusual domestic set-up will result in hilarious consequences. I'm picturing a lot of scenes where the lady goose will say something like, 'I like shopping for shoes and eating chocolate!' and then the two gay geese dads will say, 'So do we!' much like you already do with that *Will & Grace* show. Really, it writes itself.

Think of it as *Ellen* meets *My Two Dads*. Only with geese. I have no strong feeling either way regarding whether it would be best to use actual geese, puppets or have Bobcat Goldthwaite in a goose costume.

Looking forward to your speedy reply.

Best wishes,

Gid

A bit gay.

RATTLEBOX MOTH

```
SCIENTIFIC NAME..Utethesia ornatrix
CLASSIFICATION...Phylum: Arthropoda, Class: Insecta, Order: Lepidoptera,
                 Family: Arctiidae
SIZE.............33-46 mm
DISTRIBUTION.....North American east coast
DIET.............Plants
```

The most dangerous moment in any relationship comes when it's time to give each other gifts. There you are, thinking you've got some genuinely deep emotional connection, when your beloved hands you a parcel. 'Go on, open it! You're going to love it! See? It's a toy Crazy Frog! Like off the ringtone adverts. You know the ones. Ding ding ding! Aren't they just the funniest thing? Also, I got you this book about how animals have sex!' And in that one grim moment of clarity you fully appreciate how truly and terribly alone we're all destined to be.

Gift-giving in the animal world is an equally hit-and-miss affair. Balloon-fly males give their partners a little silk parcel, which she excitedly unwraps whilst he has sex with her, only to find out that it was empty all along. Luckily things are a bit better for the rattlebox moth, as he already knows full well what she wants: pyrrolizidine alkaloids. He gets these from eating rattlebox plants and then passes them to her in his semen. You're probably thinking, 'Yeah, fair enough. What girl doesn't want a few pyrrolizidine alkaloids? They're okay, but it's not exactly a trip to the Caribbean, is it?' In moth circles it's much better than that, because they give the female special **anti-spider powers**, which are the moth's main predator. It's certainly more useful than those moths that have a musical vibrating penis, which might be impressive at first, but I bet the novelty soon wears off.

The only downside for the lady moth is that she has to get **a face full of spider lick**, because it's not until the spider has started trying to eat her that he realises how bad she tastes. After that he doesn't even want to have the moth anywhere near him, so he'll cut her free from his web and shoo her away. You should note that this is about the only time in this entire book, or indeed Nature, where a male actually contributes something genuinely useful, so make the most of it.

GIRAFFE

```
SCIENTIFIC NAME..Giraffa camelopardalis
CLASSIFICATION...Phylum: Chordata, Class: Mammalia, Order: Artiodactyla,
                 Family: Giraffidae
SIZE.............5-5.5 m, up to 900 kg
DISTRIBUTION.....Africa
DIET.............Leaves, twigs
```

As far as the male giraffe is concerned, evolution has already played enough of a dirty trick by leaving him having to pump blood around his body at twice the pressure of any other mammal, just to get it up his improbable neck and into his brain. So he's not going to give the process a helping hand or put any more strain on his heart unless it really seems worthwhile.

To this end the male giraffe is only ever interested in sex if there's a good chance the female will get pregnant. If he reckons it's the right sort of time, he'll give her a nudge on the bum with his head. This usually has the effect of making her urinate. Then, like some perverse Oz Clarke, **the male laps up her urine** and swills it round his mouth for a bit. Having used his sensitive palate to detect whether or not she is fertile – 'I'm getting mostly warm giraffe wee…but with just a hint of delicious estrus chemicals' – he'll try to mount her. Except females aren't much more interested in the whole sex business themselves, so fairly often, halfway through, she'll wander off towards a nice-looking tree, leaving him almost falling over his own spindly legs.

GARTER SNAKE

SCIENTIFIC NAME..*Thamnophis sirtalis*
CLASSIFICATION...Phylum: Chordata, Class: Reptilia, Order: Squamata,
                 Suborder: Serpentes, Family: Colubridae
SIZE.............60-120 cm
DISTRIBUTION.....North America
DIET.............Slugs, snails, worms, lizards, leeches, mice

Try to think of anything more alarming than a great big ball of snakes rolling towards you. Possibly a great big ball of bears would be worse. Or a ball of sharks. But a ball of snakes is still pretty terrifying.

Snake **mating balls** – often made up of over a hundred individual snakes – occur each spring when one of the female garter snakes emerges from an underground den and releases a powerful pheromone that makes every male snake in the vicinity up sticks and rush over to pile on top of her. Assuming she doesn't get crushed to death, one of the males will eventually manage to fight off the other snakes for long enough to insert one of his two penises into her cloaca.

But there's a twist. Mating occurs just as the snakes are coming out of hibernation, which leads to another age-old problem facing boys: you want to get the girl, but you're **lazy**. And it's cold outside, and to be honest you haven't really woken up properly. All those other male garter-snake jocks seem to be getting frisky, whilst you're trying to get back to that nice dream about swallowing an entire rabbit, so what chance do you have?

The solution some of these sleepier snakes have managed to come up with is to give off the female pheromone themselves. Pretty soon he's in the middle of a mating ball, with the rest of the boys all pointlessly fighting each other tooth and nail – or snakey equivalent – to try to impregnate him. This has two advantages: 1) he's nice and snug in the centre of the ball, so he doesn't lose as much body heat, which is always important for cold-blooded reptiles, and 2) the other males all **knacker themselves out** fighting each other to get to him, so by the time our lazy hero has warmed up enough to get in the mood to go looking for a female, he's got much less competition from everybody else. Knocks all that early-bird protestant work-ethic nonsense right on the head.

GARDEN SNAIL

```
SCIENTIFIC NAME..Helix aspersa
CLASSIFICATION...Phylum: Mollusc, Class: Gastropda, Order: Pulmonata,
                 Family: Helicidae
SIZE.............2.6-3.3 cm
DISTRIBUTION.....Europe, California, Canada, Mexico, New Zealand, Chile, Argentina
DIET.............Almost any vegetation, bark
```

Garden snails have their own version of cupid, except sensibly they don't actually need a chubby gold-painted baby-man to help things along.

Being hermaphrodites, any snail can make out with any other snail. When two snails happen to meet, they go through a courtship ritual that they probably find extremely exciting and erotic, but which to the casual observer, just looks like a couple of snails circling around each other very, very slowly. It's actually a kind of sexy snail version of **laser quest**: what they're really doing is trying to get into the best position to fire – but not get hit by – a **snail love dart**.

The snail love dart doesn't actually impregnate the other snail; that's done in a separate bit of snails-squelching-their-faces-together – because the snail's penis is next to its ear – but it does affect the target snail's uterus, allowing it to store much more **snail sperm**. The snails do their best to avoid getting hit for a number of reasons: 1) there's no reproductive benefit for the target, only for the shooter, 2) having a dart fired through your belly and into your uterus isn't much fun, and 3) snails are lousy shots, so there's a good chance you'll end up getting speared through the brain.

SCORPION

```
SCIENTIFIC NAME..Buthus occitanus
CLASSIFICATION...Phylum: Arthropoda, Class: Arachnida, Order: Scorpiones,
                 Family: Buthids
SIZE.............Up to 21 cm
DISTRIBUTION.....Southern hemisphere, North America
DIET.............Insects
```

It's often said that there's a fine line between love and hate. Obviously this isn't true. They're complete opposites! A child could tell you that. But at least when you're dealing with scorpions, there is a grain of sense to this otherwise **profoundly stupid** saying.

Slightly one-note creatures, if you want to create an accurate computer simulation of a scorpion, you don't need to go much beyond:

```
10 Look sinister
20 Sting the crap out of something
30 Go to 10
```

And they're not fools. They know their reputation. They realise bikers don't have lots of scorpion tattoos to appear charming and débonnaire, and that those shadowy organisations bent on world domination aren't all called things like S.C.O.R.P.I.O. for no good reason. So when a couple meet, it's always there in the back of **their little arachnid minds** that, much as they really do want to have sex, they also wouldn't mind having a stab at killing each other, just from force of habit. This at least lends the proceedings a certain frisson.

First off, and as gingerly as possible, the scorpions grab each others' claws, because it's getting pincered to death rather than stung that they've got to watch out for, seeing as scorpions are immune to their own poison. Though sometimes the male will sting the female because it seems to calm her down a bit. What then ensues is almost always described in the scientific literature as a **'love dance'**. It actually amounts to some clumsy manoeuvring by the male in the hope that the female sits on a packet of sperm he deposited on the ground a bit earlier. Note to scorpion scientists: this is not anybody's definition of 'dancing', and if you try it on a real lady, you will get in a lot of trouble.

BED BUG

SCIENTIFIC NAME..*Cimex lectularius*
CLASSIFICATION...Phylum: Arthropoda, Class: Insecta, Order: Hemiptera,
                 Suborder: Heteroptera, Family: Cimicidae
SIZE.............4-5 mm
DISTRIBUTION.....Any temperate climate
DIET.............Blood

So far as all this sex business goes, a lot of the smallest creatures seem to be the ones that get up to the worst stuff. Possibly this is just a result of sheer biomass, ie there are so many different species of insect that statistically a few of them are bound to do something disgusting – or possibly it's because, being almost microscopic, they misguidedly think nobody is going to notice.

Bed-bug sex is problematical. Not regular 'Should I pretend to her that I work with terminally ill children?/Have I got squid ink all round my face from that Japanese restaurant?' problematical. I mean tricky on a much more fundamental level: notably, the female doesn't have any sex organs. You might think this would be an insurmountable hurdle, but it doesn't make the male bed bug miss a beat. He employs the unusual method of **traumatic insemination** – that's 'stabbing the female through her side, so that his sperm mixes with her blood' to you – with his specially-designed-for-the-purpose 'intromittent organ' – or 'a dirty great prong'.

Research suggests that males in some species of bed bug have gone one step further and developed the ploy of **injecting their sperm into the abdomens of other males**. This means that when the second male finds a female, there's a chance he'll pass on his rival's sperm as well as his own.

So. All that is going on. Right under your ear. Sleep tight, don't let the bed bugs traumatically inseminate you through your belly!

In the words of Nick Ross:
don't have nightmares.

Much less sinister
than actual clowns.

CLOWNFISH

```
SCIENTIFIC NAME..Amphiprion percula
CLASSIFICATION...Phylum: Chordata, Class: Actinopterygii, Order: Perciformes,
                 Family: Pomacentridae
SIZE.............9-12 cm
DISTRIBUTION.....Warm waters of the Pacific
DIET.............Small crustaceans
```

Back in the 1940s, disgruntled Disney animators – bored out of their skulls by the painstaking two-year process it took to make Mickey so much as scratch his nose – famously used to relieve their frustrations by inserting flash-frames of Thumper sporting a giant cock into the middle of *Bambi*.

Computers have made animation a whole lot faster, so there probably isn't an X-rated version of *Finding Nemo* floating about the Internet. This is a shame, because the title already sounds like it might be one of those **late-night, Channel Four European films** about the first awakenings of sexual identity. Which in Nemo's case turns out to be pretty ambiguous.

Clownfish live happily in groups, with the males for once, instead of being totally useless, helping to guard the eggs. But if something untoward happens to the female, like she gets fished up by the man from Petsmart and sold to some spoilt child who's just seen a Disney film, then the males don't hang about and mope. **The biggest and bossiest one changes into a girl**. He's already got some unused female genitalia stored away just in case, so it's not too much bother. If he gets fished up as well, then the next bossiest male changes into a girl and so on. This sort of behaviour *almost* makes up for the disappointment of realising that clownfish spend their lives hanging about with boring sea anemones instead of getting in and out of tiny cars.

This is all going to end in tears

HONEY BEE

```
SCIENTIFIC NAME..Apis mellifera
CLASSIFICATION...Phylum: Arthropoda, Class: Insecta, Order: Hymenoptera,
                 Family: Apidae
SIZE.............1.5 cm
DISTRIBUTION.....Native to Europe, Asia, Africa
DIET.............Nectar
```

'I never dreamed that it would turn out to be the bees!
They've always been our friends!'

Of all the stupid things in Irwin Allen's *The Swarm* – and there's a lot of stupid things in *The Swarm*, possibly more than in any other film ever made – the most stupid is how surprised Michael Caine is that the bees have finally gone a bit mental and decided to wipe us out.

   During her lifetime the queen bee will make just a couple of mating flights, and on these flights she will mate with only a handful of drones. Which means the male bee has about a **one in twenty thousand chance of having sex**.[2] It gets worse, because halfway through mating the 'lucky' bee explodes, his genitals snap off inside the queen, and he drops dead with his guts spilling out all over the place. So is it any wonder that they're always really angry? You'd be trying to take over the world as part of a killer swarm, or attack Texas, or sting Winnie the Pooh to death **if your penis had just snapped off**. Especially if some biologist kept trying to give you the 'it makes perfect sense from an evolutionary point of view, because the male's discarded genitals act as a sort of chastity belt to prevent any other bee from mating with the queen' line, which is frankly no comfort at all given the situation.

[2] Kids, if you're going to graffiti this book, crossing out 'male bee' here and writing in the name of somebody you don't like might be a good place to start.

RAT

SCIENTIFIC NAME..*Rattus rattus*
CLASSIFICATION...Phylum: Chordata, Class: Mammalia, Order: Rodentia, Family: Muridae
SIZE.............Up to 0.5 kg
DISTRIBUTION.....Worldwide
DIET............Almost anything

Starting plagues. Weeing in rivers. Making babies autistic. Rats get the blame for everything. I think it's probably all Aristotle's fault. In *The History of Animals*, he shares with us his brilliant theory of where rats come from. They don't have sex, he reckons, instead they 'spring, fully formed, from mud'. Aristotle is full of stuff like this. 'Elephants mate rear to rear', 'Men go grey at the temples and not at the back, because the back of the head doesn't contain any brain'. To be fair, Aristotle was spot on most of the time, but he gives the impression of being a bit like the man down the pub who can't ever bring himself to admit there might be something outside the scope of his knowledge, so he ends up making all sorts up. Aristotle is just the Ancient Greek version of Cliff from *Cheers* in this respect. Anyway, rats have never really recovered from this ignominious start.

Though having said all that, they don't do much to help their reputation, at least where sex is concerned. When a male rat ejaculates, he produces, along with his sperm, a kind of putty goo that sets hard and forms a genital plug in the female. It's exactly the same principle as the exploding bee genitals, but a bit less drastic. And – this is the bit all the twelve-year-old kids in the audience will rate – it's a little bit useless, because other male rats will come along and then **eat the genital plug** as a nice pre-coital energy snack.

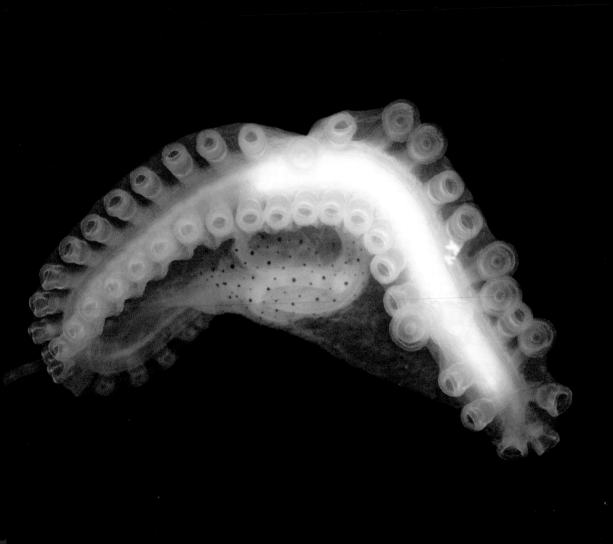

PAPER NAUTILUS (OCTOPUS)

```
SCIENTIFIC NAME..Argonauta argo
CLASSIFICATION...Phylum: Mollusca, Class: Cephalopoda, Order: Octopoda,
                 Family: Octopodidae
SIZE.............Females 20 cm, Males 4 cm
DISTRIBUTION.....Global, tropical and subtropical waters
DIET.............Small fish, crustaceans
```

Some things can really make you take a long, hard look at what you're doing with your life. Watching a sad documentary about tiny Romanian orphans. Seeing a baby lamb being born. Finding yourself typing the sentence **'detachable swimming penis'**.

Octopuses are strange in a whole lot of ways. They can squeeze through holes the size of a fifty-pence piece. If miserable, they have a tendency to eat themselves. And one recent study shows that they like nothing better than pretending to be coconuts drifting along the seabed. They're also very shy, so when an octopus finds a mate, he prefers to try to keep the whole business of having sex at a literal arm's length. Upon meeting a girl, the male produces a huge 'spermatophore', which is basically a big sperm hand grenade. He picks it up with one of his arms, gingerly reaches across and pushes it up into the female's mantle cavity (this is also what she happens to breathe through) where it then explodes. Using your hand to put your sperm up your girlfriend's nose. It's just not really very romantic.

But in one species, the paper nautilus, they take the whole aloof thing that bit further. For these creatures, the spermatophore-carrying arm actually breaks off and swims under its own power into the female. The male then begins to waste away and dies not long after. Possibly the trauma of watching his penis swim off without him has something to do with this. I don't know what happens if the penis misses and accidentally swims into a shark's mouth or something. Presumably the nautilus has a bit of a cry.

BOWERBIRD

```
SCIENTIFIC NAME..Ptilonorhynchus violaceus
CLASSIFICATION...Phylum: Chordata, Class: Aves, Order: Passeriformes,
                 Family: Ptilonorhynchidae
SIZE.............25-30 cm
DISTRIBUTION.....Australia, New Guinea
DIET.............Berries, worms
```

Right at the other end of the 'aloof'/'putting-a-bit-too-much-effort-in-to-have-any-dignity' spectrum comes the bowerbird.

It's one thing spending an afternoon trying to find the best place on your coffee table to position that book of boring French poetry so you can look all sophisticated in the unlikely event that lovely girl from accounts ever comes round your flat, but the male bowerbird spends his entire life doing this. Except it's not boring books of French poetry and coffee tables, it's little collections of anything from berries to bottle tops to parrot feathers to coloured moss. And they're incredibly anal about where each item should go so that they best complement each other: if you put a blue pebble in the middle of their pile of green pebbles, **they throw a fit**. Some species even paint the walls of their bower with pieces of bark dipped into a mixture of bird spit and leaf matter just to set everything off nicely. The only thing that interrupts a bowerbird from piling up his dead snails and twigs and all the other crap he collects is when he takes a moment to visit another male's bower to steal his best pieces and generally mess the place up.

The female (who lives in a separate, regular nest) will occasionally come by to inspect the place, whilst the males sit there trying to look nonchalant – 'Those are my beetle carapaces. A lot of bowerbirds seem to have beetle carapaces nowadays, but I've been into them for ages. Ah, I see you've noticed that bat's skull. There's a funny story behind that...' – knowing that she'll only mate with whoever is judged to have the prettiest collection.

So how did this improbable **sex-for-art-galleries** system evolve? The best theory is that by putting all that effort into his display, the male is unconsciously showing off how much genetic fitness he has to spare. It's a case of 'look, I've just spent two whole days fannying about with all this nonsense, so clearly I don't have anything more important to worry about, like disease or recessive genes or that sort of thing.' You could try using this as an excuse when your girlfriend discovers that collection of hand-painted *Lord of the Rings* role-playing figurines, but I wouldn't bank on it working.

Doing stupid things to impress girls.

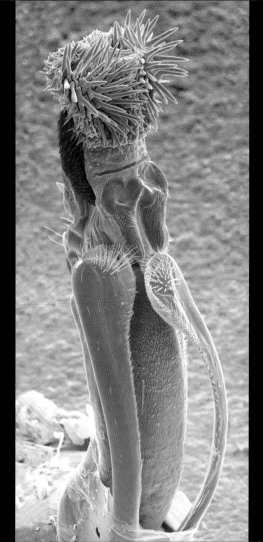

Jesus. H. Christ.

BEAN WEEVIL

SCIENTIFIC NAME..*Callosobruchus maculatus*
CLASSIFICATION...Phylum: Arthropoda, Class: Insecta, Order: Coleoptera,
                 Family: Bruchidae
SIZE.............0.6 cm
DISTRIBUTION.....The Tropics
DIET.............Beans

Creatures never get together to indulge in the creature-sex equivalent of the 'You think you were poor? When I was a kid, I had to eat coal,' conversation. Why not? **Mrs Weevil would always win**. Mrs Weevil has a terrible time of it. Think about Mrs Weevil and her lot in life next time you're watching couples on *Trisha* moan at each other for getting Jaffa Cake crumbs in the bed.

Because, in an attempt to put the female off any ideas of sleeping around, Mr Weevil has evolved **the worst penis in the world**. It looks like something Torquemada designed just after a heretic punched his favourite puppy to death. Covered in **spikes and barbs**, the male weevil's penis does a pretty effective job of making the female weevil one of the creatures least prone to have casual sex – because even a single sexual encounter can drastically reduce her lifespan. The only upside is that Mrs Weevil has evolved an enormously powerful set of hind legs, so at least she can give her unappealing lover a really good kicking.

COMMON CHIMP

SCIENTIFIC NAME..*Pan troglodytes*
CLASSIFICATION...Phylum: Chordata, Class: Mammalia, Order: Primates,
                 Family: Hominidae
SIZE............1-1.7 m, Females 25-50 kg, Males 34-70 kg
DISTRIBUTION.....African rainforest
DIET............Fruit, seeds, insects

When they're not blasting off into space to fight those evil communist dogs, or busy taking over the world in a **bleak dystopian future**, chimps like nothing better than having themselves lots of very noisy sex. This is why if you gave an infinite number of chimps an infinite number of typewriters, you'd just end up with a lot of broken typewriters that the chimps have dismantled and turned into sex toys.

In the 1920s one man, **Serge Voronof**, was so impressed by the chimpanzee's prodigious sex drive that, in a series of appallingly misconceived operations, he attempted to graft chimp testicles on to elderly men in order to 'reinvigorate' them. The results were not great. But why are chimps, especially female chimps – who have been observed copulating with half a dozen different males in a matter of minutes – so rampant in the first place? A big cause is probably infanticide. If a male chimp defeats a higher-status chimp in a fight to gain access to his females, then one of the first things he's inclined to do is kill any baby chimps who happen to be about. This has the effect of bringing the females back into heat. The 'obfuscation' theory is that by sleeping around so much, the female chimps are helping to prevent too much of this sort of thing, because none of the males can be really sure that it's not their own children they're trying to knock off.

It should be made clear that, despite our near-identical genetic make-up, bashing a human female's kids to bits against a tree rarely gets her in the mood for sex.

LYNX SPIDER

SCIENTIFIC NAME..*Oxyopes salticus*
CLASSIFICATION...Phylum: Arthropoda, Class: Arachnida, Order: Araneae,
                 Family: Oxyopidae
SIZE.............4-10 mm
DISTRIBUTION.....North America, Central America, West Indies
DIET.............Insects

When biologists talk about 'parallel-evolution', what they're referring to is when two different organisms happen to share common features, not because of a mutual ancestor, but because they have both adapted to their environments in a similar fashion. A famous example of this is a certain nameless delicate-nosed kids' TV presenter and spiders. Both are creepy. Both have poisonous fangs that can kill a child. And both have a real knack for getting caught indulging in dubious sexual practices at bondage parlours.

To be fair to the male spider, his fetish is at least born out of necessity. Female spiders tend to be huge and frightening, whilst the male is usually stunted in comparison. This makes sex a dangerous business. The solution of the male lynx spider is to tie up the female in **little silk ropes**. Thankfully this as far as the bondage goes; you won't come across any spiders wearing miniature eight-eyed gas masks. Only once he's convinced she's secure does he dare to actually mate with her. Lacking a penis, the male must first pick up a handful of the sperm he secretes and then place it inside the female using his feelers.

What's strange – I mean, it's all strange, **it's spider sex**, but what's particularly strange – is that the female seems to just be playing along with the whole situation, because as soon as they've done their thing she breaks from her bonds without any difficulty (the only exception being the species *Ancylometes bogotensis*, where the male does a proper, thorough job of trussing her up, and it takes her a good couple of hours to escape). It's thought that the male's pheromone-soaked silk might provide the female with a constant chemical reminder of her suitor's identity, so that she doesn't just forget what's going on and who he is halfway through the sex and mistake him for a tasty spider-shaped muffin.

ATELOPUS FROG

SCIENTIFIC NAME..*Atelopus chiriquiensis*
CLASSIFICATION...Phylum: Chordata, Class: Amphibia, Order: Anura, Family: Bufonidae
SIZE.............3-5 cm
DISTRIBUTION.....South America
DIET.............Insects

AWKWARD POST-COITAL CONVERSATION #1:

'Well, that was nice.'

'It was nice, wasn't it?'

'So… I don't mean to be rude, but I notice you're still sitting on my back.'

'Yes. I'm going to stay here for a little while, if that's okay.'

'Oh. Right.' [Pregnant pause.] 'So, when you say "a little while"…'

'Six months.'

'Pardon? Did you say six months? You're going to go on lying on top of me for six months?'

'Pretty much.'

'Is this one of those clever ploys to prevent me from having sex with anybody else?'

'Might be. I'm not saying. So, this weather we've been having. Bit rainy.'

'We're in a rainforest.'

Six months. Even **self-proclaimed tantric-sex guru** Sting doesn't have that kind of staying power. Along with that barnacle-penis business it's enough to make you feel inadequate, assuming you're the kind of person who can be made to feel sexually threatened by a crustacean and a frog. Though because the frog just squirts some semen on to the female's back at the same time as she releases her eggs, they're not technically copulating for all this time. But the twig insect does, for a couple of months. You can still feel inadequate over that if you want.

You might want to read a
book or something, because
this could take some time.

PORCUPINE

SCIENTIFIC NAME.. *Hystrix cristata*
CLASSIFICATION... Phylum: Chordata, Class: Mammalia, Order: Rodentia,
                  Family: Hystricognathi
SIZE............. 60-90 cm, 6-16 kg
DISTRIBUTION..... Asia, the Americas, Africa, Europe
DIET............. Leaves, bark

I'm contractually obliged to cover porcupines, but that doesn't mean I'm going to do the old 'How do porcupines have sex?' chestnut, so if that's the kind of cheap gag you're after you can skip to dolphins.

To the male porcupine's dismay the female is only interested in sex for about four hours a year. This means that for 364 and a half days he just has to make do with rubbing himself against a stick. But come the big afternoon, the male stands proudly on his hind legs and waddles towards the female, showing off his erection. Once he's about six feet away he drenches the object of his desires with **huge spurts of porcupine wee**. Unbelievably some female porcupines don't seem to find this the sexiest thing in the world, in which case she'll just shake off the urine and give him a punch, leaving the male to disappointedly go and find solace with his stick again. But if the mood has finally taken her, then she'll raise her tail, exposing her quill-free underside, reverse towards the male and let him mount her from behind. He quite often keeps his paws above his head **to avoid getting stabbed**. After they've finished, and seemingly keen to make up for lost time, the female will go looking for another male, and another after that, and as many as she can fit in to her brief period of feeling sexy. And then, as quickly as it all started, it's over, and the male wanders back home to stare wistfully at his calendar for another twelve months.

AMERICAN BURYING BEETLE

```
SCIENTIFIC NAME..Nicrophorus americanus
CLASSIFICATION...Phylum: Arthropoda, Class: Insecta, Order: Coleoptera,
                 Family: Silphidae
SIZE.............3 cm
DISTRIBUTION.....North America
DIET.............Small carrion
```

The only trouble with goths is knowing that beneath all the eyeliner and tassles and slightly septic noserings, they're probably quite nice middle-class kids. And most of them don't get up to anything more gothic than watching Buffy marathons or practising their glowering face in the mirror.

Goth girls! Prove your satanic mettle by telling goth boys that you'll only have sex with them if they bring you **an actual rotting corpse**. This is proper gothic behaviour, and it's what the female American burying beetle demands of her man. She'll only put out once he's found her the carcass of a small bird or rodent. After they've had sex next to it, they then set to work pulling the feathers, or fur, or what-have-you off the body. Then they spit on it for a while. Then they bury it. Finally they dig a little tunnel for her to lay some eggs in. And when the kids are born, the happy couple will both help their children eat the corpse, often by regurgitating it for them. It makes you feel all broody, doesn't it?

PRAYING MANTIS

SCIENTIFIC NAME.. *Mantis sphodromantis*
CLASSIFICATION... Phylum: Arthropoda, Class: Insecta, Order: Mantodea,
                  Family: Mantidae
SIZE............. 8-12 cm
DISTRIBUTION..... Global
DIET............. Insects

You're a single-parent mother mantis and your kids want to know about where babies come from. Unfortunately you can't just avoid the issue like any normal parent in the hope that the little wretches find out for themselves off of the Internet. Instead, you have to sit them down on your scaly mantis knee and tell them the appalling truth. Which goes something like this:

'When a mummy mantis and a daddy mantis love each other very much, the mummy mantis will give the daddy **a special sort of hug**. Whilst they're doing their special hug, the mummy mantis starts to eat daddy's eyes. Then she eats the rest of his head. Eventually she chews through a bunch of nerve endings in his brain stem that inhibit his libido. This makes daddy go into a series of satisfying sexual spasms. You see, it turns out daddy is actually a lot better at sex once he's dead. Anyhow, the mummy mantis goes on eating daddy, leaving his genitals, still working away, till last. I hope this has answered your question. Now go play with your sister.'

The upside is that the kiddie mantises will almost certainly never bring up any awkward subjects ever again.

SIDE-BLOTCHED LIZARD

```
SCIENTIFIC NAME..Uta stansburiana
CLASSIFICATION...Phylum: Chordata, Class: Reptilia, Order: Squamata,
                 Family: Phrynosomatidae
SIZE.............5-6 cm
DISTRIBUTION.....Western North America
DIET............Insects, scorpions, spiders
```

An exciting nature game for two players, in case you ever find yourself stuck in the otherwise achingly dull Utah.

You will need:
ONE LADY LIZARD
ONE YELLOW SIDE-BLOTCHED MAN LIZARD
ONE ORANGE SIDE-BLOTCHED MAN LIZARD
ONE BLUE SIDE-BLOTCHED MAN LIZARD.

Rules:
Aim of the game is to have sex with the lady lizard. Hold a man lizard of your choice behind your back. Count to three. Reveal your lizard.

Each lizard has a different characteristic: **orange** males are fiercely aggressive and guard large territories; **blue** males are aggressive, but less so than the oranges, and guard smaller territories; the **yellow** males aren't aggressive and don't guard any territory, but they are **sneaky**.

The more aggressive oranges will beat blues and get the girl in any confrontation. But oranges are too busy scrapping and patrolling about the place to notice the unobtrusive yellows, who can sneak past and mate with the lady lizard. Blues, having a smaller territory to guard, do notice the yellows and can fight them off.

The game works on exactly the same principle as **scissors-paper-stone**, except it's much more exciting, because you use lizards instead of making shapes with your hand.

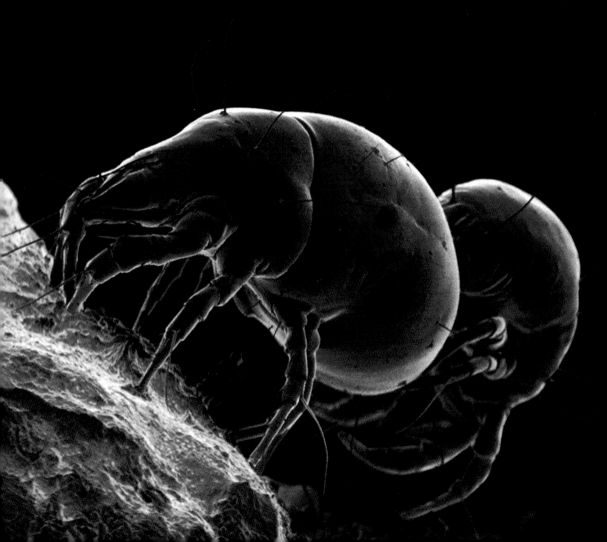

MITE

```
SCIENTIFIC NAME..Pyemotes herfsi
CLASSIFICATION...Phylum: Arthropoda, Class: Arachnida, Order: Acari,
                 Family: Pyemotidae
SIZE.............Microscopic
DISTRIBUTION.....Native to Europe, also found in North America
DIET.............Moth larvae
```

The problem of playing up the parallels between humans and the rest of the animal kingdom for laughs is that sooner or later you run into something like the moth mite. And then you find yourself having to make jokes about what is basically incest and child abuse. Before you know it you've got people from Aldershot painting 'Paediatrician scum!' on the side of your house.

When a female moth mite gives birth to a son, he doesn't head off to make his way in the world and find a moth of his own to go and bother. Instead, **like an even smaller version of Ronnie Corbett** in *Sorry*, he stays at home with his mum. Here, he waits for a baby sister to be born, actually going so far as to act obstetrician and help to deliver her (he'll only help deliver his sisters, though – he ignores any brothers). And as soon as he's pulled her out he forcibly mates with her.[3] She then flies off, possibly a bit put out by the pace of life nowadays, whilst he waits for the next sibling to pop out.

Obviously just as you can't say anything useful about human morality from what animals get up to, you also can't make judgements about animals based on what we think is right or wrong. But even so. It's his SISTER. And she's about ONE SECOND OLD. Did I mention that he eats bits of his mother whilst he's at it? Honestly, moth mites!

[3] The reason all this inbreeding doesn't lead to vacant-eyed, banjo-playing *Deliverance*-style moth mites is because of something called paternal genome elimination. Put as simply as possible, this means the male moth mite's father's genes are wiped out early in his development, so he only has the genes from his mum. This means he can't be concealing any recessive genes to pass on to his daughters – if he's got any genetic problems, they'll be up front and obvious.

DUGONG (SEA COW)

```
SCIENTIFIC NAME..Dugong dugong
CLASSIFICATION...Phylum: Chordata, Class: Mammalia, Order: Sirenia,
                 Family: Dugongidae
SIZE.............2-3 m, 300-400 kg approx
DISTRIBUTION.....East African/Australian coast
DIET.............Sea grasses
```

People from the olden days were stupid. I can back up this sweeping statement simply by pointing your attention to the adjacent photograph. This is a dugong. In no way do I wish to knock the now sadly endangered dugongs, who are great animals, if a little clumsy. But according to a number of sources, the **ancient myth of mermaids** arose from sailors seeing dugongs, and their close cousins the manatees. Look at it. Look. At. It. 'Not as beautiful as they are painted,' observed Christopher Columbus, master of the massive understatement. There are even tales of fishermen hauling the unfortunate creatures on to their boats in order to have sex with them. However long you've been at sea, and however lonely you may feel, this is not acceptable behaviour.

When they're not being molested by us, dugongs occasionally form **leks**. Lekking is when you have a group of boys all marking out a specific territory that the girls then come to visit in order to have sex with the best man there. The actual lekking behaviour varies from animal to animal. Cichlid fish show off their pointless ability to build sandcastles. Peacocks just do their best to make a right racket. And dugongs, or at least the ones in Shark Bay, Australia, do a sort of workout, involving 'sit-ups' and 'belly-ups'. Even though they're not really the right shape to be able to kiss their own biceps, this is still possibly the only activity that could make them look more ridiculous than they already do.

Hubba hubba.

Idiots.

DOLPHINS

```
SCIENTIFIC NAME..Stenella frontalis
CLASSIFICATION...Phylum: Chordata, Class: Mammalia, Order: Cetacea,
                 Family: Delphinidae
SIZE.............50-200 kg
DISTRIBUTION.....Global, temperate waters
DIET.............Fish, some crustaceans
```

Ah, dolphins! Always leaping about in front of fractals or big yin-and-yang symbols on those motivational posters, beloved of middle managers and New Age cretins everywhere.

When during the Gulf War the US Army's specially trained dolphins just swam off instead of looking for mines like they were supposed to, it was generally held up as final proof of how smart they really were. I have a different theory, which is that they simply forgot all about it and decided to have sex instead. Though not necessarily because they saw a passing lady dolphin. Maybe they glimpsed a nice-looking turtle. Or the shapely curve of an eel. Or a rock. Because to us, a rock is a rock. But to the dolphin, a rock is An Exciting Sexual Challenge.

The male dolphin's penis is an impressively dextrous **prehensile swivelling effort**, which they use to explore things under the water. But it's as if they think, 'Well, I've used my penis to explore this object, it's turned out to be the bottom of a boat, but what the hell? Whilst my penis happens to be there, I might as well try to have sex with it anyhow. Here we go!' They are the least discriminating animals in the world, and they seem to find everything equally attractive. It was probably dolphins who started that dugongs-are-just-like-lovely-mermaids rumour. Dolphin lovers put this all down to **healthy vim and pep**. Which is one expression for it. Another would be 'dolphins are slappers'.

More dubiously, males also indulge in what look very much like gang rapes, where a group of them will herd a female and then try to forcibly mate with her. Only this doesn't sit so well with all the panpipe, henna tattoo, woolly Eastern philosophy stuff, so people tend to gloss over that in favour of the bouncing-a-ball-on-their-nose malarkey.

MORMON CRICKET

```
SCIENTIFIC NAME..Anabrus simplex
CLASSIFICATION...Phylum: Arthropoda, Class: Insecta, Order: Orthoptera,
                 Family: Tettigoniidae
SIZE.............3-4 g
DISTRIBUTION.....North America
DIET.............Various plants
```

Biologists have a hard time explaining how the current Western male preoccupation with stick-thin skeleton ladies might have come about. Given that being just 12% or so underweight can render a woman infertile, it doesn't make much sense from either an evolutionary or any other kind of perspective.

The Mormon cricket also has a bit of a weight fixation, but it's a much healthier one. When looking for a mate, he'll insist on **weighing the female before having sex with her**. This doesn't earn him the slap upside the head you might think he deserves, because in the Mormon-cricket world, and contrary to what you find with most other creatures, it's the women who are fighting each other to have sex with the men. The male will only mate with the biggest girl he can find. And once he makes his choice, he'll have sex whilst she eats the huge spermatophore that he produces. This isn't the usual easy-to-produce blob of sperm; it's packed with water and nutrients and represents both a big effort on the male's part and a full-on banquet for the girls – which is why the topsy-turvy competition has evolved in the first place. The boy gets his big girl, which helps the girls get even fatter, and **everybody's happy**.

I have tried to fight back the urge to do a Sir-Clicks-a-Lot joke, but I'm weak. Sir-Clicks-a-Lot! Because it's like Sir-Mix-a-Lot, who prefers the larger lady, but with crickets, who make clicking sounds, you see? Oh, suit yourself.

BLUEGILL

SCIENTIFIC NAME..*Lepomis macrochirus*
CLASSIFICATION...Phylum: Chordata, Class: Actinopterygii, Order: Perciformes,
                 Family: Centrarchidae
SIZE.............25 cm, 1.5 lbs
DISTRIBUTION.....North America
DIET.............Insects

It's not just the cruel world of high school that has a rigidly enforced social structure of jocks and geeks. A number of underwater creatures have adopted the same model, the best example being the bluegill. The large, over-developed **jock male** finds a girl and shows her to his nest. He then refuses to let her out of his sight. Just like in humans, these bluegill jocks appear to be pretty stupid, often attacking creatures much bigger than themselves – even human divers – in the mistaken belief that they are after their woman, like a kind of angry aquatic red-faced Phil Mitchell. But they have good cause to be so overprotective, because there's not one but two types of **nerd bluegill** trying to get their girl.

The first is a smaller type of male who simply tries to dart into the nest and dump some sperm onto any eggs that happen to be lying about. The second kind is more devious. He's disguised to look exactly like a girl. He swims along to the nest, all brazen and flirtatious: 'Don't mind me, I'm just a girl. You know what I like to have sex with? Boys. Can't get enough of them. Big muscles, all that kind of thing. Really does it for me,' and the jock bluegill puffs out his chest and gives himself a pat on the back for being so irresistible. Meanwhile the new ladyboy is happily fertilising all the eggs. And by the time the jock has his **Des Lynam moment of clarity**, it's too late: he's left guarding a nest full of somebody else's kids.

RED-BACKED SPIDER

SCIENTIFIC NAME	*Latrodectus hasselti*
CLASSIFICATION	Phylum: Arthropoda, Class: Arachnida, Order: Araneae, Family: Theridiidae
SIZE	5 cm
DISTRIBUTION	Australia
DIET	Insects

There's a pig's head in the butcher's down the road from me that has a cardboard speech bubble emerging from its mouth which says, 'Buy me!' Whether the speech bubble was written by a specially bred pig that was eager to be eaten and had been taught to write and use scissors before its death, or whether it has been put there by the butcher in an unusual and possibly misguided attempt at marketing is a matter that can only be answered with further research from pigologists. However, there is an animal that definitely does want to be eaten – the male red-backed spider. I know we've seen how other spiders and scorpions and mantises can wind up as elevenses, but that's only if they badly mess things up. Most of time they get away with it and have an exciting story to tell their friends afterwards. But the red-backed spider does his best to try to ensure his mate devours him by doing **an acrobatic backflip right into her mouth halfway through having sex**. If she tries to spit him out, he gets really upset and attempts to clamber back in.

The reason he's so keen to be her tea is because if she's just eaten a plate full of tasty spider abdomen, the female is much less likely to bother having sex again anytime soon. You know how it is after a nice big meal, you don't really feel like doing anything except having a bit of a lie down. So the male spider is both contributing something to her health – and in turn that of his kids – but also preventing any other male coming along to remove his sperm. Apparently the third Spiderman film is going to 'deal with some of the issues that come with being Spiderman which we haven't seen yet'. You can only hope this means we'll have scenes of Tobey Maguire trying to crawl inside a surprised-looking Kirsten Dunst's mouth.

WHIPTAIL LIZARD

SCIENTIFIC NAME..*Cnemidophorus uniparens*
CLASSIFICATION...Phylum: Chordata, Class: Reptilia, Order: Squamata, Family: Teiidae
SIZE.............15-23 cm
DISTRIBUTION.....California, Arizona
DIET.............Insects

The position of men often seems precarious. Testosterone slowly poisons us, and our stunned Y-chromosome is rapidly falling to bits. We're one collective *no-really,-I-just-think-of-you-like-a-brother* away from mass extinction.

Some species have taken that plunge already. Whiptail lizards used to have men, now they don't, and it's as simple as that. The females all immaculately conceive, giving birth to perfect clones of themselves. Though perhaps as an overhang from their more sexual past, the lady lizards still engage in bouts of **pseudocopulation**. That is, one of them pretends to be the man lizard and mounts the female. They go through all the motions of sex, bumping happily away, even though nothing is actually going on down there. Then they switch around. For some unknown reason they still seem to need to do this in order to stimulate egg production.

The trouble is, clones aren't that great. I know they sound all **shiny and futuristic**. And you're probably thinking how useful it would be to have a clone of your own, so that you could hatch those schemes where you win a marathon by having the clone take over halfway through. But in the long term almost every species that has tried giving up sex tends to go extinct after a few hundred generations. Cloning is certainly less bother than finding a partner, impressing them and convincing them to have sex with you, but without the regular recombination of our genes you just end up standing still whilst the rest of the world – ie diseases and that – overtake you. It's like Madonna: if she'd kept those big bushy eyebrows she had back in the 80s, would she have had a twenty-year career? She would not. She had to reinvent herself on an almost annual basis just to keep up. In the end, clones quickly become the equivalent of some mad old retired colonel, waving his copy of the *Telegraph* and shouting about how not enough stuff is made out of bakelite nowadays.

MACAQUE

SCIENTIFIC NAME..*Macaca mulatta*
CLASSIFICATION...Phylum: Chordata, Class: Mammalia, Order: Primates,
                 Family: Cercopithecidae
SIZE.............6-10 kg, 50 cm
DISTRIBUTION.....India, East Asia
DIET.............Fruits, seeds

If there's one thing worse than dolphins, it's our own species' **stupid concealed ovulation**. It might have been useful for our ancestors: firstly for clouding paternity issues enough to stop cavemen bashing other cavemen's kid's brains in, and secondly by encouraging dads via regular sex to stay at home and help out about the place. But it leaves us modern humans floundering about in a deeply confusing world, saying things like, 'Oh, yes, obviously I knew this was a business meeting rather than a date. I didn't misinterpret the signs at all. Why did I hire this string quartet to play romantic songs next to our table? I just thought it would help us go through the, uh, flowcharts.'

Much better is the monkey system to show you're in the mood by having your bum swell up and turn bright red. In an experiment carried out at Duke University Medical Center, they discovered that macaques would give up treats – in the form of free orange juice – just to look at pictures of lady macaques' asses. Whilst the panda business had shown that animals might get off on porn, this was the first time it had been demonstrated that some creatures would actually be **prepared to pay for it**. The macaques also gave up their juice supply for the chance to look at pictures of high-ranking males from their social group, suggesting that there is also a gap in the market for a celebrity-based *Heat*-style magazine aimed at monkeys. 'Exclusive pictures of the SKINNIEST capuchins yet!' – that kind of thing.

Yes, we have no
banana slug penises.

BANANA SLUG

SCIENTIFIC NAME..*Ariolimax dolichyphallus*
CLASSIFICATION...Phylum: Mollusca, Class: Gastropoda, Order: Palmonda,
                 Family: Arionidae
SIZE.............25 cm
DISTRIBUTION.....North American Pacific coast
DIET............Leaves, animal droppings

When the producers of *Star Trek* want some kind of space worm and are feeling particularly cheap, it's the scary-looking banana slug that they turn to, *Ariolimax dolichyphallus*.

The reason science involves so much Latin is because it's a lot easier to get funding for 'Studies of the Reproductive Habits of *Ariolimax dolichyphallus*' rather than what it translates as, which is roughly 'We'd like to poke about with something called the big penis slug. That's what we've named it, because it has a big penis, and it's a slug.' There's another bit of Latin for you, because Banana Slugs have the distinguishing habit of engaging in **appophalation**, better known as **'chewing each others' penises off'**. They do this either as a result of getting stuck, because their penises are simply too large – at ten inches almost as big as their body – or because of the hermaphrodite thing again, where it's in each slug's interest to have the other one act the female, and eating their penis is a pretty sure-fire way of achieving this.

You see? You thought you weren't going to learn anything useful from this book. Now you can try looking worldly and sophisticated by dropping 'appophalation' into dinner-party conversation.

Super disco breakin'

MANAKIN

SCIENTIFIC NAME..*Manacus manacus*
CLASSIFICATION...Phylum: Chordata, Class: Aves, Order: Passeriformes, Family: Pipridae
SIZE............10-12 cm
DISTRIBUTION.....Central and South America
DIET.............Insects, fruits, berries

If you can't dance, the endless flood of magazine articles with titles like 'How Dancing Ability Exactly Reflects Sexual Performance and Stamina in Bed as Proven to be Scientific Fact by Actual Scientists' gets a little depressing. This is the myth that the female manakin has bought into, and as a result the male manakin has evolved to become the best dancer in the world. They're like Kevin Bacon, Patrick Swayze and John Travolta rolled into one, except obviously anything with the composite face of Bacon/Swayze/Travolta would be a **terrible abomination against nature**, and manakins are actually quite attractive little birds.

I cannot emphasise enough how fantastic the manakin dance is. Their trademark move is the might-sound-a-bit-familiar **walking backwards along a twig whilst it looks like their feet are walking forwards**. And during their display they flap their wings at 160 times a second, which is faster than a hummingbird. Try flapping your arms 160 times a second. It's pretty tiring. Manakins lose up to 10% of their body fat doing their dance routines. Once again it's an example of sexual selection making the boys handicap themselves in order to demonstrate just how fit they are.

And the picky manakin female wants more for her viewing pleasure than just one boy dancing about. She demands to see two birds carrying out a choreographed dance routine together, even though she'll only sleep with one of them. So each dominant male – that's basically the one who owns the log where they do all their performances – has to take on an apprentice, who he teaches all the right moves to, keeping him loyal with vague promises of 'one day, lad, all this will be yours.'

I have a dream of remaking *Electric Boogaloo* with manakins, where the plucky little birds win a dance competition to stop greedy developers knocking down their forest. I would knit them little hip-hop hats and jackets. But for now it must remain just that, a dream.

HYENA

SCIENTIFIC NAME..*Crocuta crocuta*
CLASSIFICATION...Phylum: Chordata, Class: Mammalia, Order: Carnivora,
                 Family: Hyaenidae
SIZE.............85 cm, 65 kg
DISTRIBUTION.....Africa, Asia
DIET.............Wildebeest, zebras etc

MICHAEL GRADE, BBC

Dear Michael,

Still no word from you on my geese sitcom? Probably you are busy building the puppets, assuming that's the option you went for. In the meantime I have had a further fantastic idea for a comedy show based around animals.

My latest work features **a female hyena** travelling back in time to the Victorian era (the actual time-travel explanation can perhaps be a cross-over with your excellent *Billie Piper Goes to Space* programme). This series will work on the Fish-Out-of-Water principle of comedy that was so successful in both *Crocodile Dundee* and *Crocodile Dundee 2*.

For you see, females in the Victorian era were incredibly genteel and ladylike. This is the exact opposite of hyena females! Hyena females are bigger, stronger and more aggressive than hyena males, and they even have a sort of penis, which is actually an enlarged clitoris they can erect at will. When having sex, the male **has to insert his penis into her pseudopenis**, which is no easy task. And they actually greet each other by waggling their sex organs at one another. Obviously this will cause a storm in the buttoned-down Victorian times. I envisage the show to be something like *M∗A∗S∗H*, with the occasional unfunny story that deals with sad issues, such as the way 10% of hyena females die in childbirth, because giving birth through a penis is even more difficult than having intercourse through it. I think that episode might even win us a Bafta.

Regards,

Gid

RHINOCEROS

```
SCIENTIFIC NAME..Diceros bicornis
CLASSIFICATION...Phylum: Chordata, Class: Mammalia, Order: Perissodactyla,
                 Family: Rhinocerotidae
SIZE.............Up to 2.2 tonnes
DISTRIBUTION.....Africa, Asia
DIET.............Plants, foliage
```

Hey, ladies! Why not try the no-nonsense rhino approach to finding out whether your man is worth putting out for? It's easy! Stand at one end of your hallway and get him to stand at the other. Now run really fast at each other until you both collide with a **big clanging head-butt**. To get the full effect, it helps if you each weigh about three tonnes and can get up to speeds of 30 miles per hour. If your man is now lying unconscious on the floor bleeding badly from a nasty head wound, he's probably not worth bothering with. But if he's still on his feet after you've repeated the process a few times, then he could just be the one.

This kind of effort is why male rhinos living in safari parks seem to prefer having sex with passing Renault Lagunas than with actual females.

ARGENTINE LAKE DUCK

```
SCIENTIFIC NAME..Oxyura vittata
CLASSIFICATION...Phylum: Chordata, Class: Aves, Order: Anseriformes, Family: Anatidae
SIZE.............640 g, 40 cm
DISTRIBUTION.....Argentine lakes, obviously
DIET.............Seeds, plant remains, small invertebrates
```

The easiest way to understand all scientific concepts is to ask the question 'If this scientific concept was a character from the popular sitcom *Friends*, which character would it be?' **Natural selection** is best thought of as **Ross** – because it results in sensible things like eyes and lungs. Whereas **sexual selection** is best thought of as **Joey** – because it results in stupid things like peacocks' tails and bower birds' bowers and the Argentine lake duck's penis.[4]

Most birds don't have penises at all. With alarming ease they've happily dispensed of them in favour of the slightly rubbish 'brief pressing together of genital orifices'. But not only has the Argentine lake duck kept his penis, he's become embroiled in a daft sexual arms race. This has left a not very big duck with an unfeasibly outsized organ of up to 41 centimetres in length. The sexual selection theory would put this down to female choice – a preference for longer penises that has become magnified over several generations. There's a limit to how far this process this can go, because otherwise it will all get so out of proportion to the rest of his body that he'll end up sinking to the bottom of his Argentine lake under the sheer weight of the thing. Sexual selection can work a lot faster than natural selection, but it's the latter that's always going to have the last laugh.[5]

Though having said that, scientists at the University of Alaska think sexual selection may only be partly responsible for the duck's freakish physiology. It could also be a result of intense sperm competition. Or it may be used in aggressive displays to intimidate other rival males. And, brilliantly, **it's been suggested that they use their penises to lasso reluctant females**. No research has been carried out to discover whether they go cattle rustling with them too.

[4] Seeing as a lot of the results of sexual selection are not just stupid but also kind of funny, a case could be made for it being like Chandler before he got boring and fat, but I don't want to confuse the issue.

[5] At this point you will have noticed the Joey/Ross analogy breaks down altogether. Ross never gets any sort of laugh, except in that episode where he had the glowing teeth.

Of course, none of these theories matter, because you're not even reading this. You're still just staring in slack jawed disbelief at the picture opposite. Close your mouth before you start to get dribble everywhere.

TORTOISE

```
SCIENTIFIC NAME..Geochelone nigra
CLASSIFICATION...Phylum: Chordata, Class: Reptilia, Order: Testudines,
                 Family: Testudinidae
SIZE.............5 cm - 2 m
DISTRIBUTION.....Global
DIET.............Plants, grasses
```

A while back there was an 'And Finally' news story doing the rounds about how British tortoises were getting too fat. Overfed by adoring owners when they were babies, their soft shells would bulge out at the bottom, and when they solidified a few months later, the poor things wouldn't be able to get about, because their legs couldn't reach the floor. The reason it made the news was because a vet came up with the brilliant solution of **jacking up the rear of the tortoises with sets of Lego wheels**. This worked a treat and opened up endless possibilities of turning your pet tortoise into a half tortoise/half Lego star destroyer and getting them to have battles in the back garden.

The fat tortoise issue also comes into play at mating time, when the females tend to swell up. Unfortunately this means she can't be quite so choosy as she might like, because she becomes a bit too big for her shell. Her usual response to unwanted male attention is to simply pull her rear in, so the male tortoise can't get to any of her bits. But when she's fat all he has to do is walk round the front and bite her on the nose. This makes her pull her head in, which in turn makes her ass stick out. So he trots round the back of her again and remounts.

One aspect of tortoise sex they never showed on *Blue Peter* is the disturbing noise they make. It's an eerily human groaning. Not very realistic groaning, though – more like exaggerated, badly acted porno-film groaning. Darwin complained that the noise of Galapagos giant tortoises going at it kept him awake at night. But given that Galapagos giant tortoises don't reach sexual maturity until they're about **forty years old**, it's fair enough that they should be pretty excited to have finally got the whole business underway.

STARFISH

SCIENTIFIC NAME.. *Asterias rubens*
CLASSIFICATION... Phylum: Echinodermata, Class: Asteroidea, Order: Forcipulatida, Family: Asteriidae
SIZE............. 6-30 cm
DISTRIBUTION..... Global
DIET............. Plankton, insects, molluscs

There was a kid at my school called Graham who had one eye bigger than the other. Nobody really talked to him much. Then one day in class we learnt that starfish could make new starfish by wrapping one arm around a rock, and another around something equally immovable and then having a tug-of-war until it succeeded in pulling itself apart. Hey presto, you've got two starfish.[6] Next break-time there was Graham with his arms wrapped round a couple of drainpipes, gurning with effort.

'What are you *doing*?' we asked.

'I'm going to pull myself in two and make a new friend!' replied Graham.

As cries for help go, you can't get much more obvious than that. And on some level we all knew this and realised that we were at least in part responsible. But in a world as sensitive to social rules as the school playground, making out you were a starfish and trying to rip yourself in two wasn't going to do you any favours. Especially when you've already got one eye bigger than the other. The only positive thing you can take from this story is that at least we weren't taught about how starfish sometimes eat things by **pushing their stomachs outside of their body**. That could have got messy.

[6] Actually this kind of asexual fission is pretty rare: starfish mostly reproduce by the usual boring method of shooting out big clouds of eggs or sperm and hoping they all meet up somewhere, so Graham ensured his continued social alienation as a result of not double-checking the facts.

BUMBLEBEE EELWORM (NEMATODE)

SCIENTIFIC NAME.. *Sphaerularia bombi*
CLASSIFICATION...Phylum: Nematoda, Class: Secernentea, Order: Tylenchida,
Suborder: Spaerularina
SIZE............Microscopic
DISTRIBUTION.....Europe, North America
DIET............Parasite

I like bumblebees, so I'm automatically predisposed not to like anything that gives them a hard time. Such as the nematode *Sphaerularia bombi*. There's a good chance you've never heard of nematodes, and an even better chance you're going to wish it had stayed that way. Unappealing microscopic little worm things, a male and a female will crawl about in the mud until they find each other, at which point he'll quickly ejaculate and die. She then crawls off to find her ideal accommodation, which happens to be the guts of a queen bumblebee. Once she's burrowed her way in she'll start to make herself comfortable. Then the really loopy part begins. Her sex organs begin to grow. And grow. And grow. It's like something from a disturbing, creature-orientated spam email: 'Nematodes! Would you like to **grow your vagina and uterus to FOUR THOUSAND times the size of the rest of your body**? Click here for some hot bumblebee host bodies!'

Eventually the nematode's now gigantic sex organs detach and lead an independent existence for a bit, whilst the rest of the body withers away and dies. Finally the little baby nematodes pop out and get passed in the bumblebee faeces. And the whole miserable cycle begins again. It was a can of nematodes that made it back to Earth after the space-shuttle disaster. Which is a pity, because I think outer space is the best place for them.

I know it's a little difficult to keep up that 'glorious Nature in all her wonder' attitude when confronted by this kind of thing, but you just have to bite your lip and remember that it's exactly the same blind algorithmic process gave us these miniature monsters as gave us **Jennifer Garner's cheekbones**/George Clooney's teeth. Swings and roundabouts. Bear with it, because next we've got everyone's favourite...

Here's a nice picture of a bumblebee, because you really don't want to see what's going on inside it.

PENGUIN

```
SCIENTIFIC NAME..Pygoscelis adeliae
CLASSIFICATION...Phylum: Chordata, Class: Aves, Order: Sphenisciformes,
                 Family: Spheniscidae
SIZE.............45-65 cm
DISTRIBUTION.....Antarctica, southern hemisphere
DIET.............Krill, fish, squid
```

What you hear on Pingu:

PINGU: Waaahhh!
ROBBIE THE SEAL: Wah waahhh.
PINGU: Waah wah waaaahhh!
ROBBIE THE SEAL: Waaahh.
PINGU: Wah waaahh waah wah.
ROBBIE THE SEAL: Waahh waahhh.

What Pingu is probably saying:

PINGU: Hello, Robbie.
ROBBIE THE SEAL: Hello, Pingu. That's a very shiny rock you've got there.
PINGU: Yes. I read that some female penguins – even those already part of an otherwise monogamous couple – are happy to prostitute themselves in return for big shiny rocks. [Pingu waggles his flipper suggestively.]
ROBBIE THE SEAL: Lady penguins do have that reputation. They seem to have an almost insatiable love of rocks.
PINGU [sadly]: Of course, the problem for us penguins is that, even if I meet a girl, we're both shaped like milk bottles.

ROBBIE THE SEAL: So?

PINGU: Have you ever tried to balance a milk bottle on top of another milk bottle? It's pretty tricky. Lady penguins lie down and try to hold really still, but even then us males tend to just fall off.

ROBBIE THE SEAL: That must be a little embarrassing.

PINGU: It is.

LION

```
SCIENTIFIC NAME..Panthera leo
CLASSIFICATION...Phylum: Chordata, Class: Mammalia, Order: Carnivora, Family: Felidae
SIZE.............Females 118 kg, Males 250 kg
DISTRIBUTION.....African plains
DIET.............Wildebeest, zebras etc
```

It's the oldest philosophical debate in the world. Which are best: lions or bears?

The bear argument goes something like this: 'Bears are best because we are always getting up to stuff. Also, we're good at dancing. And the songs in *The Jungle Book* – main creature, a bear – are great, whereas the songs in *The Lion King* – main creature, a lion – are soulless tripe. Lions on the other hand, or male lions at least, are lazy good-for-nothings who let their women do all the hunting whilst they just take naps.'

'Rubbish,' say the lions. 'Yes, male lions do spend most of their time trying to decide which is the best rock to have a sleep on, but there's a really good justification for this. Namely, the lionesses being **insatiable nyphomaniacs**, demanding that the male has sex up to **150 times a day for five days running**. Besides which, we're Kings of the Jungle. That's not the kind of title they just hand out with breakfast cereal, you know.'

'Maybe so,' say the bears, 'but we have a football team named after us. And a type of stock market! Match that! Raagh!'

To which the lion replies, 'You're a bunch of idiots. It's a "bull" market that's the good kind of stock market. A "bear" market is when shares are going down.'

'Damn it,' say the bears, and they wander off sulkily to go and mess up some bees.

ELEPHANT

```
SCIENTIFIC NAME..Loxodonta africana
CLASSIFICATION...Phylum: Chordata, Class: Mammalia, Order: Proboscidea,
                 Family: Elephantidae
SIZE.............4 m, 7500 kg
DISTRIBUTION.....Africa, Asia
DIET.............Grass, leaves, twigs, roots
```

The Elephant Man undoubtedly had a hard time of it trying to get by in the less than enlightened midst of Victorian society. On the upside, had he been an **actual elephant**, things could have been even worse. At least when John Merrick met an attractive girl, he just looked a bit forlorn and felt miserable about how it could never work out – he didn't go insane, start weeing himself on an almost constant basis and find his penis had turned green. Such is the lot of the male elephant.

For years naturalists thought that elephants suffered from 'green penis disease', until they realised it was a side-effect of musth. Musth is a corruption of the Urdu word for 'intoxicated', and that's exactly how the males behave during the mating season: they bellow at other males, their testosterone levels shoot up to fifty times higher than is normal, turning them hyper-aggressive, and then there's the weeing business. A bull elephant loses about **350 litres of urine a day** while he's in musth. Faced with a lover this irritable and incontinent, you might expect the females to not be all that interested, but in his favour, even though it's turned green, the male's prehensile penis is the biggest of any land animal.[7]

It has to be, because female's vagina isn't just beneath her tail, like you might expect, but much further round under her – this means that the male has to lean back on two legs in order to get to it, which is a good thing, seeing as the last thing she needs is **four tonnes of excited elephant** putting most of his weight on her back. Once the sex is over, the female elephant's relatives are comically proud of her conquest. In what's known as 'mating pandemonium', her entire family – sisters, mother, grannies and everything – will rush over and give her a big collective 'That's my girl!' by shouting, trumpeting and defecating everywhere.

[7] As are his testicles (though these are safely tucked away inside him, because elephants aren't particularly high off the ground, and there's spiky plants and that about the place).

My other car is a beetle.

PSEUDOSCORPION

SCIENTIFIC NAME.. *Cordylochernos scorpioides*
CLASSIFICATION... Phylum: Arthropoda, Class: Arachnida, Order: Pseudoscorpionida,
Family: Cheliferidae
SIZE............. 2 mm
DISTRIBUTION..... Global
DIET............. Mites

When you're called something like a 'pseudoscorpion', you're automatically starting off on the back foot. Everyone's reaction to you is going to be, 'So you're not *actually* a scorpion? You're a sort of pretend scorpion? What kind of an occupation is that?' The way to counter this slur, and to impress girls whilst you're at it, is to get yourself some **classy-looking transport**. Which is why most of the time you'll find male pseudoscorpions riding about on the backs of harlequin beetles or the like, cruising around the neighbourhood logs and making regular stops to pick up, and then have sex with, the extremely promiscuous females. They're a regular little insect Ferris Bueller.

It sounds like irresponsible behaviour, but in fact they show an admirable concern for proper health and safety by making themselves tiny silk seat belts. I know it's a pity they don't take the analogy any further by having **harlequin-beetle drag races**, or suffering mid-life crises and suddenly trading in their old beetle for a beetle that looks a bit more like an enormous penis, but you can't have everything.

Damn you,
lack of sperm
competition,
for making
me this way!

GORILLA

SCIENTIFIC NAME..*Gorilla gorilla*
CLASSIFICATION...Phylum: Chordata, Class: Mammalia, Order: Primates,
                 Family: Hominidae
SIZE............Females 70-140 kg, Males 135-275 kg
DISTRIBUTION.....Central Africa
DIET............Leaves, shoots, stems

AWKWARD POST-COITAL CONVERSATION #2:

KONG: You've gone a bit quiet.
FAY WRAY: Oh, you know. I was just thinking.
KONG: It's my tiny penis, isn't it?
FAY WRAY: What? No! Of course not.
[There's an embarrassed silence. Kong swats at a biplane.]
FAY WRAY: It's just it is quite little.
KONG: It's not my fault! It's my social structure!
FAY WRAY: Of course it is. Your social structure is to blame. I understand.
KONG: Listen, Fay. Gorilla males guard isolated harems of females. Any competition between males is limited by the fact that we really don't want to mess with each other, because look at us, we're terrifying. I could pull David Attenborough's legs off without even breaking a sweat. You only need big reproductive organs when you've got a sperm competition going on.
FAY WRAY: You're right. It all makes sense. Even so. **Two inches**.

FRUIT FLY

```
SCIENTIFIC NAME..Drosophila bifurca
CLASSIFICATION...Phylum: Arthropoda, Class: Insecta, Order: Diptera,
                 Family: Drosophilidae
SIZE.............2.5 mm - 4 mm
DISTRIBUTION.....All tropical regions
DIET.............Rotting fruit
```

Contrary to what you might have read in books, mind control is usually a terrible way to try to pick up girls. More often than not you'll just wind up with a slap, or at least several angry questions along the lines of 'Why are you talking in that spooky voice? And what's with all this dangling a pocket watch in front of my face? Are you trying to hypnotise me into sleeping with you again? I've warned you about this before.' It rarely ends well.

Male fruit flies sensibly leave the mind control to their sperm, which contains a whole **cocktail of chemicals** that can help make his mate more compliant and lower her libido when it comes to making out with any other male. Though it's amazing that fruit flies ever find the time to have sex, given that they spend most of their lives being bombarded with gamma rays and the like by curious scientists eager to see if they'll grow legs out of where they'd normally have eyes.

Despite repeated and energetic attempts, mainly by teenagers, a human male has never managed to produce an individual sperm bigger than his own body. Which is a relief, because it would probably look like **a great big translucent ghost** and scare the wits out of everybody. But this is what the male fly does, opting for the quality-over-quantity principle with a single massive five centimetre-long sperm, twenty times longer than himself, and it takes him three days to make it. Which is another reason it's a good thing human males don't do the same, because anything that took three whole days to make would probably lead to a lot of self-congratulation and a sperm-equivalent to those World's Biggest Marrow competitions.

Like a dirty underwater
version of the Three
Musketeers

FLATWORM

SCIENTIFIC NAME.. *Pseudoceros bifurcus*
CLASSIFICATION... Phylum: Platyhelminthes, Class: Turbellaria, Order: Polycladida
SIZE............. 6 cm
DISTRIBUTION..... West Pacific
DIET............. Ascidians

Homo sapiens' most important innovation is not fire, or wheels, or even those NASA space pens that can write upside down and on greaseproof paper. It's the ability to understand metaphor.

'But those pens are amazing!' you cry. 'How can some **lousy metaphors** be better than space pens? So we can read a bunch of angsty teenage poetry and know that the "creeping spiders" are really symbols of "feeling very miserable"? Big deal.'

But actually metaphors save us from getting in a lot of bother. When human males want to compete with each other (read: compare penis size), they can do it through the medium of running quite fast, or the size of their iPod's hard-disk, or their ability to eat an entire mixed grill, or any number of ways. Flatworms don't have metaphors, so instead they literally do battle it out with their penises. It's a variation on the hermaphrodite snail love-dart theme, and it's called **penis fencing**, which sounds like something Errol Flynn invented in between takes of *Captain Blood*, but is actually a serious and violent encounter. Each flatworm has two sharp penises coming out of where their mouths should be, and they use these during long duels to try to rip holes in each other. This can go on for quite some time, until eventually one of them scores a hit, injecting sperm into any part of the other flatworm's body, which then migrates towards the ovaries and gets the loser pregnant.

STALK-EYED FLY

SCIENTIFIC NAME..*Cyrtodiopsis dalmanni*
CLASSIFICATION...Phylum: Arthropoda, Class: Insecta, Order: Diptera,
                 Family: Diopsidae
SIZE............4 mm
DISTRIBUTION.....South East Asia
DIET............Leaves

If somebody tells you that 'you have nice eyes', you might as well give up right there. Because what they're really saying is 'I have examined your entire face and have been unable to find a single feature worth praising, so instead I am complimenting you on your eyes, which is meaningless, seeing as how everybody has nice eyes. Eyes are just eyes for goodness' sakes.'

But the female stalk-eyed fly really means it, and if your eyes aren't on **stalks set further apart than the entire length of the rest of your body,** she won't even give you a second look. The competing males line up and measure eye-stalks against each other and she mates with whoever has the longest. The loser gracefully backs off, cheering himself up with the thought that at least he can still grab his stalks with his feelers and pretend to be **flossing his brain**, which is a gag that never gets tired in stalk-eyed fly circles.

Of course, in humans, eyes being too close together simply shows that you're carrying the 'not to be trusted' or 'thieving' gene.

FRIGATE BIRD

```
SCIENTIFIC NAME..Fregata magnificens
CLASSIFICATION...Phylum: Chordata, Class: Aves, Order: Pelecaniformes,
                 Family: Fregatidae
SIZE.............2.5 m wingspan, 1.5 kg
DISTRIBUTION.....Over tropical oceans
DIET.............Fish, small turtles
```

One thing to bear in mind when looking at how animals have sex is that you shouldn't draw too many conclusions from it. Just because something does or does not occur in nature doesn't make it right or wrong so far as humans are concerned. For example – there's a terrible documentary always doing the rounds on cable about men who cheerfully admit that they can only get aroused when in a room full of gigantic balloons. And I'm pretty sure this is deeply, unforgivably wrong, even though the female frigate bird shares the exact same fetish.

A sort of fat feathery Richard Branson, it can take the male frigate bird up to twenty minutes to inflate the huge red balloon they have dangling from their necks. The females come along to inspect who has the biggest and **best and shiniest balloon**, and will then have sex with whoever she judges that to be. During sex the male will, rather sweetly, put his wings over the female's eyes to make sure she doesn't get distracted by any passing male with a nicer-looking balloon.

The problem is that sharp beaks and big balloons are a really stupid combination, especially when you haven't evolved to a level where you can do that clever sellotape/pin trick. Rival males will do their best to go about popping other males' balloons. This is disastrous for the victims, because once your balloon has burst, that's it, you're never going to get a girl. You're doomed to spend the rest of your life looking a bit sad with a **shrivelled up piece of skin hanging off your chin**. As with so many things, the Fresh Prince probably said it best: 'I don't want to burst your bubble, but girls ain't nothing but trouble.'

PIG

```
SCIENTIFIC NAME..Sus scrofa
CLASSIFICATION...Phylum: Chordata, Class: Mammalia, Order: Artiodactyla,
                 Family: Suidae
SIZE.............90-150 cm
DISTRIBUTION.....Global
DIET.............Almost anything
```

Budapest is a fine city, and it has lots of worthwhile things to visit. Yet for some reason on a recent trip there, a friend and I became obsessed with seeing an exhibit mentioned in the *Time Out Guide to Budapest*: an aerosol in the National Agricultural Museum inexplicably marked 'PIG SEX'. What could it be? Was it a more accurately named Hungarian version of Lynx? Was it pheromones that would make you irresistible to any pig? Or just a regular can of spray paint farmers would use to draw sexy, fulsome lips on a lady pig? Unfortunately when we arrived, the agricultural museum had been temporarily replaced by an exhibition of '**Every Kinder Egg Toy Ever Made**'. So we never did discover what was in that aerosol, though it was nice to see a full set of Teeny Terrapins.

I'm sorry that anecdote didn't really go anywhere, but here's what I *can* tell you about pig sex. Pigs have a long slender corkscrew penis with a hole in the side, out of which they ejaculate an **entire pint of semen** (the last few ounces of which is like a thick, lumpy custard, to keep it all from falling out again). Like a lot of creatures' penises, this particularly odd shape may have something to do with preventing hybridisation. The more unique the penis the more likely it will only fit one kind of vagina, which stops different species trying to have sex with each other. This is a good thing, because hybrids are generally useless and infertile. The exception to this would be something that was half eagle and half shark, which would *rule*.

ANGLER FISH

SCIENTIFIC NAME..*Edriolychnus schmidti*
CLASSIFICATION...Phylum: Chordata, Class: Osteichthyes, Order: Lophiiformes,
                 Family: Eeratidae
SIZE............Females 20 cm, Males 3 cm
DISTRIBUTION.....Atlantic, Pacific Indian Ocean
DIET.............Fish, larger invertebrates

Finally a creature that mates for life. Isn't that sweet! Romantic little fishes! With lovely big eyes!

Haven't you learnt anything by now?

It's obvious just by looking at them that angler fish are going to get up to something terrible. If you gave a child a set of crayons and told them to draw a nightmare, odds are the end result would be a picture of an angler fish. The best you can say of the things is that they look exactly like a creature who lives in the spooky black depths of the ocean *should* look.

It's this bleak environment that's to blame for the angler fish's unconventional sex life. Even with the **natty little headlamp** the girls have stuck to the top of them, it's still incredibly difficult to meet each other down there in the gloom, so when the (tiny) male finally finds a (huge) female, he does his alarmingly literal best to never let her go. Not with any of that bringing her gifts/showing off his amazing colours/doing a funny dance business those other creatures get up to. The male angler fish has the much more direct approach of taking a big bite out of the side of her, and then hanging on for dear life. He hangs on for so long his jaw actually fuses with his mate's skin. This gets round the problem of not having a digestive tract of his own, because he starts to live parasitically off her blood supply. Over time he wastes away until eventually he's just a pair of gonads stuck to her side, supplying sperm as and when required.

And on an **entirely unrelated note**, Kate Moss is still going out with him from the Libertines.

EEL

```
SCIENTIFIC NAME..Conger Conger
CLASSIFICATION...Phylum: Chordata, Class: Actinopterygii, Order: Anguilliformes,
                 Family: Congridae
SIZE.............Up to 2.7 m
DISTRIBUTION.....Europe, North America
DIET............Insects, crustaceans, snails
```

How many great scientific thinkers does it take to come up with a lot of old rubbish about eel sex? The answer is lots. To summarise, here are a few of the theories about how eels reproduce, courtesy of Aristotle, Pliny, Albertus Magnus and the like:

Eels come from mud, just like those rats.

Eels come, in a vague and unspecified way that we're not going to be drawn on, from beetles.

Eels are made out of a special type of **dewdrop**.

Eels are formed from the bits of old skin that fall off other eels.

The dewdrop one is quite good, but the rest are like something Stan Lee might have invented as the origin story for one of the less impressive *Marvel* superheroes when he was having a particularly lazy afternoon. I'm not saying this to have a go at some of the finest minds in human history. Well, all right, I am a bit, but mostly it just illustrates how difficult biology can be. People were trying to figure out how eels reproduced for a couple of thousand years, looking at them under big magnifying glasses, injecting them with stuff, throwing them at each other in frustration, but it wasn't until 1897 that they even started to get anywhere close to the truth.

It turns out that the reason you never find a baby eel is that the setting for reproduction has got to be just right. No matter where they come from, the rivers of Europe or North America, they all swim thousands of miles for an **annual secret sex conference** in the middle of the Sargasso Sea. But it's still not known how or why they go all that way. Presumably that particular bit of water has some special sexy quality that isn't remotely apparent to the rest of us.

And on an entirely unrelated note, Kate Moss…

ZEBRA FINCH

```
SCIENTIFIC NAME..Taeniopygia guttata
CLASSIFICATION...Phylum: Chordata, Class: Aves, Order: Passeriformes,
                 Family: Estrildidae
SIZE.............9-12 cm
DISTRIBUTION.....Australia
DIET.............Seeds, insects
```

If you're told enough times by enough different magazines that Brad Pitt or Paris Hilton are incredibly beautiful people, you start to accept it as a fact. And it's only years later when all the fuss has died down that you'll wake up with a start and think, 'Hang on! He has a head like a sort of boiled potato! And she not only has the dead, soulless eyes of a sociopath, but is also kind of plate-faced at the same time.'

Such is the power of **cultural conditioning**. It works in the rest of the animal kingdom too, as recently demonstrated by an experiment carried out at the snappily titled Institute for Integrative Bird Behaviour Studies in Virginia. They put two cages next to each other, one containing a boy and a girl zebra finch and the other containing just boys. These cages were left in full view of a group of females kept in a third cage. Later on, when given a choice, the females preferred mating with the boys they had already seen apparently 'paired up' with a girl. This works on the same principle as carrying around a picture of Jennifer Garner and Ben Affleck with your face Photoshopped over Ben Affleck's face, which you can then show to girls and say 'That's me when I was dating Jennifer.' No male zebra finch has yet evolved even a rudimentary knowledge of Photoshop, so they must make do with crayons and glue.

The easily influenced
zebra finch.

SEA HORSE

SCIENTIFIC NAME..*Hippocampus breviceps*
CLASSIFICATION...Phylum: Chordata, Class: Actinopterygii, Order: Gasterosteiformes,
                 Family: Syngnathidae
SIZE.............16-18 cm
DISTRIBUTION.....Eastern Atlantic, Mediterranean, Black Sea
DIET.............Small invertebrates

MICHAEL GRADE, BBC

Mikey,

I can't say I'm not disappointed that you haven't written back asking to commission my time-travelling hyena sitcom. But in retrospect I can see how it's the kind of subject matter that might be deemed a little too near the knuckle for a primetime audience.

With this in mind I am enclosing my latest work, which I am tentatively calling *You Must Be the Husband!* I'm aware there was a similarly titled comedy starring Tim Brooke-Taylor some years back, but that was rubbish, and I think you'll find it lacked the exclamation mark, which makes all the difference.

My new show features a family of sea horses, a species in which, as I'm sure you already know, it's the male who gets pregnant. That's not exactly true, the female actually deposits eggs into his pouch using her tiny penis-like organ, which he then carries about until the baby sea horses are ready to pop out. But you can imagine the scope this scenario offers for hilarious role-reversal comedy. If you can't imagine it then I certainly can't be bothered to give you examples, but I would suggest you watch the film *Junior*, which deals with many of the same issues. Indeed, if there was any chance of getting Danny Devito to reprise his role and play the pregnant sea horse's diminutive friend, that would be great.

Best wishes,
Gid

GIANT SQUID

```
SCIENTIFIC NAME..Architeuthis dux
CLASSIFICATION...Phylum: Mollusca, Class: Cephalopoda, Order: Teuthida,
                 Family: Architeuthidae
SIZE.............5-20 m
DISTRIBUTION.....Worldwide
DIET............Fish, other squid
```

The female giant squid makes her suitors jump through a fair few hoops before she puts out. First off she sets her potential suitors a task. The competing males have to find a sperm whale. Then they take it in turns to wrap their arms round the back of the whale and do their best to ride it like a sort of **oceanic bucking bronco**. The giant squid who manages to stay on the longest is afterwards invited by the female to share a mating dance. This involves the male squid making as many **obscene gestures with his tentacles** as he can, whilst the lady squid pretends to look shocked. Finally the couple retreat into an underwater den made out of old bits of crab and pull a little seaweed curtain shut behind them to preserve their modesty.

All right, I made that one up, because nobody has ever seen a giant squid alive, much less having sex. Even so, given what everything else gets up to, would you really want to bet against it?

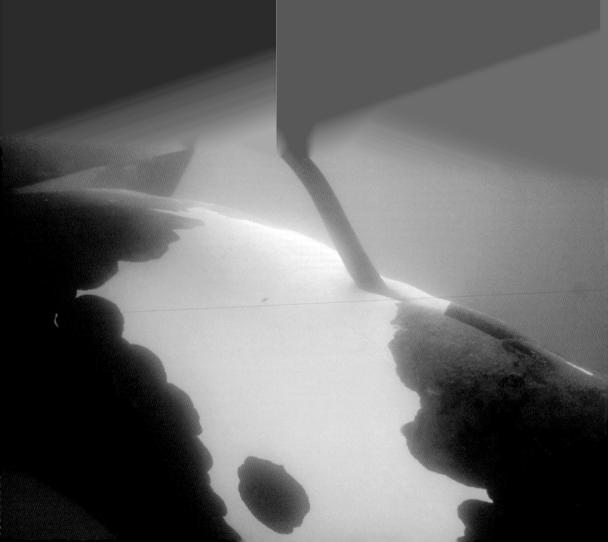

GREY WHALE

SCIENTIFIC NAME ..*Eschrichtius robustus*
CLASSIFICATION ...Phylum: Chordata, Class: Mammalia, Order: Cetacea,
                  Family: Eschrichtiidae
SIZE.............13 m, 15-40 tonnes
DISTRIBUTION.....North Pacific
DIET.............Small crustaceans

Boy meets girl. Boy really likes girl. Girl meets another boy who plays squash and has a BMW and is probably called something like 'Jean-Philippe'. At least in our species boy number one can slink home, write a couple of poems about emotions, and then go and key Jean-Philippe's stupid flash car when nobody's looking.

Grey-whale sex also often starts with two boys chasing the same girl. But when the female grey whale finally chooses her mate, the loser doesn't swim away in defeat. Basic Newtonian physics coupled with weighing lots and being really slippery, makes whale sex difficult. So the unsuccessful male has to help support the female, **like a big blubbery headboard**, whilst the other male mates with her. If he's not feeling lousy enough already, trying to pretend that he can't hear what's going on, he's also got a chin covered in barnacles, waving their outsized penises about, having sex all over the shop, just to rub in the fact that he's not getting any.

Of course the barnacles might win the penis-size-relative-to-body-mass competition, but it's whales that take the outright biggest penis in the world award,[8] so at least the grey whale can comfort himself with this thought. It's been suggested that whale penises might account for some of the ancient reports of **sea serpents**. Which is a fantastic boast to be able to make. 'Yes, I tend to scare boats off with it. That sea serpent, the one who sank the yacht? That was my *penis*.'

DRAGONFLY

```
SCIENTIFIC NAME..Aeshna cyanea
CLASSIFICATION...Phylum: Arthropoda, Class: Hexapoda, Order: Odonata,
                 Family: Aeshnidae
SIZE.............2-9 cm
DISTRIBUTION.....Global
DIET.............Mosquitoes, midges, small insects
```

If you get stuck in an argument with some Gateshead-college creationist type who, after they've finished trying to make out that dinosaurs were knocking around **just last week**, then start arguing in favour of intelligent design, try throwing dragonfly sex in their face.

Any 'design' that's gone on in regard to the way in which dragonflies mate is design carried out by a committee of imbeciles:

'Hey, let's put the sperm-making thing at the end. But we should have it so he has to masturbate first, pick up the sperm and then move it to under his belly. And give him some **sex-clamps** to grab the female by the neck. We should put her sex organs right at the tip of her body.'

'But then won't she be in entirely the wrong position to actually have sex?'

'Erm, yes. But we can have it so she kind of loops her whole body round so that it touches where his sperm is. Like in a big 'O' shape. Then let's get them to fly about in this ridiculous position, bumping into flowers and things.'

'That's brilliant. I can't believe we managed to design dragonfly morphology AND the DeLorean in half an hour. We might as well knock off early and go to lunch.'

SEA HARE

```
SCIENTIFIC NAME..Aplysia punctata
CLASSIFICATION...Phylum: Mollusca, Class: Gastropoda, Order: Anaspidea,
                 Family: Aplysiidae
SIZE.............Up to 30cm
DISTRIBUTION.....North-east Atlantic, Mediterranean
DIET.............Algae
```

It's a mystery how some of these creatures have wound up with the names that they've got. You can try squinting at sea hares for as long as you like, viewing them from any angle you might care to think of, but they still don't look anything like underwater hares. Even if you stuck a ball of cotton wool on their back and put a pair of deeley boppers on them for ears, they still wouldn't look anything like hares. They look like big aquatic slugs, because that's what they are.

So I propose a new name for the sea hare, which is **the amazing sex conga slug**. Who looks to have a fair claim for the most fun had by any creature ever. As you already know, slugs are hermaphrodites, and the amazing sex conga slug takes full advantage of this by forming **mating chains**. The slug at the front of the chain acts as a female, the slug at the back acts as a male, and all the slugs in between are both sexes at once. And once they're all locked together like this, they happily spend the afternoon shuffling along the seabed making all the other animals jealous at what an obviously brilliant time they're having.

If you feel strongly about the official adoption of the amazing sex conga slug as the sea hare's new name, why not write to:

The International Code of Zoological Nomenclature
c/o Natural History Museum
Cromwell Road
London
SW7 5BD

and get the dolphin's name changed to something stupid while you're at it.

SPOON WORM

SCIENTIFIC NAME..*Bonellia viridius*
CLASSIFICATION...Phylum: Annelida, Class: Echiuria, Family: Bonellidae
SIZE.............Female up to 2 m, Male 1-2 mm
DISTRIBUTION.....North Atlantic, Mediterranean, Red Sea
DIET.............Plankton

Angler fish don't hold the record for most useless male in the world only because of the even more sorry-looking spoon worm. When spoon worms are born, they face an immediate and important choice. If they settle on a bit of seabed where there doesn't happen to be any females about, then they develop into a female themselves, in the hope that some other spoon worms will come floating by later. But if the baby worm smells a nearby female, who basically consist of a blob the size of a **ping-pong ball and a gigantic two-metre-long tongue**, it heads right for her. Having landed on her tongue the baby turns into a male – except this doesn't really involve much effort. He never grows bigger than a few millimetres in size, and after a couple of days he decides to move from the tongue, up the female's nose and into a specially prepared chamber in the spoon worm's uterus called the *androecium*. This is where he spends the rest of his life. Eventually the female might get half a dozen of the tiny dwarf males living inside her, occasionally **gobbing up big dollops of sperm** out of their mouths but otherwise not really troubling themselves to get up to much.

If you're feeling lonely and you're keen to recreate the spoon-worm experience, why not try sticking some jelly babies up your nose, and when anybody asks what you're doing, just say happily, 'They're my new husbands!' This won't lead you to any deeper understanding of nature, but it's a good way to find out who your real friends are.

ALBATROSS

SCIENTIFIC NAME	*Diomeda exulans*
CLASSIFICATION	Phylum: Chordata, Class: Aves, Order: Procellariiformes, Family: Diomedeidae
SIZE	Wingspan up to 4 m, 8-12 kg
DISTRIBUTION	Circumpolar
DIET	Cephalopods

A favourite line of biologists is to say things like, 'And now I'd like to discuss the most DEVIANT and SHOCKING type of mating strategy in all the animal kingdom…' (Excited pause.) '… Monogamy!' And then they sit back with a smug grin going 'Ah! You expected me to say the penis-chewing thing, or something like that. You see what I did there?'

This is obviously just annoying, but it is true that proper monogamy is almost vanishingly rare in nature. So as a reward for you wading neck deep through all the weevil penises, and the traumatic inseminations, and the incest and the cannibalism, I'm giving you the albatross. When they meet they'll mate for life – which can be decades – and both parents help to bring up their young. Most importantly they **don't do anything** more exciting sexually than a blink-and-you'll-miss-it pressing their holes together.

Let's hear it for albatrosses!

APPENDIX

Some approximate penis lengths

Whale	10 ft
Elephant	5 ft
Giraffe	3 ft
Argentine lake duck	16 in
Pig	15 in
Banana Slug	9 in
Human	6 in
Gorilla	2 in

Animals Zeus manifested himself as in order to have sex, and who it was that he then had sex with

Swan	Leda
Bull	Europa
Eagle	Ganymede
A bedraggled cuckoo	Hera

Bits of animals that are used as aphrodisiacs, even though they will not make a blind bit of difference to your sex drive

Deer antlers
Tiger-penis soup
Gound-up rhino horn
Whale sick
Monkey skin

Number of pelvic thrusts made by various male primates per sexual encounter

Goeldi's marmoset	20 – 40 thrusts
Night monkey	3 – 4 thrusts
Greater bush baby	5 – 6 thrusts
Stump-tailed macaque	1 – 170 thrusts
Olive baboon	6 thrusts
Common chimpanzee	3 – 30 thrusts

(source: *Primate Societies*, Smuts et al, eds)

Creatures that men are pictured rescuing women from on the covers of 1950s men's adventure magazines

Wild boar	*True Men*
Octopus	*Bold Men*
Giant Snake	*Peril* (the All Man's Magazine)
Spider Monkeys	*Man's Life*
Gorilla	*Man's Adventure*

Some insect copulation times

Damselfly	24 minutes
Scorpionfly	65 minutes
Lovebug	50 hours
Soapberry bug	11 days
Twig insect	2 months

Rubbish pun-based Pepe Le Pew cartoon titles

Odor-able Kitty (1945)
Past Perfumance (1955)
Two Scents Worth (1955)
Heaven Scent (1956)
A Scent of the Matterhorn (1961)

Sexual dimorphism

Spoon worm	Female 200,000 times bigger than the male
Blanket octopus	Female 40,000 times bigger than the male
Angler fish	Female 20 times bigger than the male
Spectacled bear	Male 2.5 times bigger than the female
Orangutan	Male 2 times bigger than the female
Leopard	Male 1.5 times bigger than the female

Films in which apes or monkeys fall in love with human females

Title	Ape	Actress
Balaoo the Demon Baboon	Balaoo	Madeleine Grandjean
Nabonga	Nabonga	Julie London
Untamed Mistress	Nameless gorillas	Jacqueline Fontaine
Max, Mon Amour	Max	Charlotte Rampling
There's Something About Mary	Ben Stiller	Cameron Diaz

Items my dog tried unsuccessfully to have sex with

A draught excluder

My cousin's pet tortoise

A plate-glass window

My Millennium Falcon

An Amstrad word processor

ACKNOWLEDGEMENTS

Mostly, this is all Mark Rusher's fault. But Helen Garnons-Williams, Kelly Falconer and Claire Paterson aren't entirely blameless, either. I would also like to give a big shout out to my mum, Brigid, to Rob Adey, Chloe Brown, Sam Brown and Rich Murkin. Some of these people actually helped. Some of them just said things like, 'Christ on a stick, you're not still writing that stupid animal sex book, are you?' But thanks anyway.

A special mention to Mark Skarschewski, for his help with the photos.

Argentine lake duck image courtesy of Kevin G. McCracken, "McCracken, K.G., R. E. Wilson, P.J. McCracken and K.P. Johnson, 2001. 'Are ducks impressed by drakes' display'. *Nature,* 413:128"

Paper Nautilus image © Rudie Kuiter/oceanwideimages.com

Spoonworm image from www.UWPhoto.no © Rudolf Svenson

Rattlebox moth © Thomas Eisner, Cornell University

Stalk-eyed fly © Minden Pictures

Bean weevil penis, (C. maculatus) © Andrew Syred, Sheffield University/Science Photo Library

And obviously the actual scientists who wrote the books and papers that I've glibly misrepresented here are owed the biggest debt of gratitude. Please don't write in pointing out all the mistakes.

and

THE PIRATES! IN AN ADVENTURE WITH WHALING
ISBN: 0 297 84901 8 // £7.99

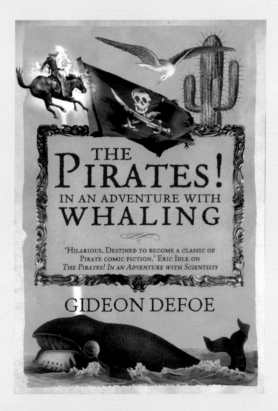